Raven Neurology Review
Clinical Neurology for
the Medical Student Clerkship

Paul D. Johnson, MD

Copyright © 2017 A Kindness of Ravens Printing Press, LLC

Denver, CO

All rights reserved.

ISBN: 1535262931
ISBN-13: 978-1535262934

DEDICATION

To my wife, Emily, who provides encouragement and support.

CONTENTS

1	Neuroanatomy	Pg 1
2	Exam Pearls	Pg 27
3	Neuroimaging basics	Pg 44
4	Neuroimaging cases	Pg 58
5	Stroke	Pg 74
5	Epilepsy	Pg 100
6	Headache	Pg 126
7	Sleep Disorders	Pg 148
8	Movement	Pg 160
9	Neuromuscular	Pg 184
10	Dementia	Pg 214
11	Immunology	Pg 230
12	Behavioral	Pg 240
13	Neurosurgery	Pg 256
14	Imaging Atlas	Pg 280
15	Case Index	Pg 303
16	Index	Pg 307

Raven Neurology Review

NOTICE

This book is intended to introduce and reinforce important clinical information in general neurology. It is specifically written for medical students in the neurology clerkship and as a study guide for the NBME neurology shelf exam and USMLE step 2. This book should not be used in isolation to make medical management and treatment decisions. Although every effort has been made to ensure the content is up-to-date and as accurate as possible, the author cannot be held legally responsible for any clinical decisions made solely based on the content of this book.

ACKNOWLEDGMENTS

I owe a great deal of thanks to the many individuals who assisted in reviewing this book and providing important feedback. Key individuals include, but are not limited to Dana Coutts PA-C, Darcy O'Banion ARNP, Allison Walczyk ARNP, Dayna Cardinal MS NP-C, Anastacia Wall PA-C, Anna Krumpe ARNP, Andrew Wolf MD, Carolyn Foster, John Foster MD, Jason Kam MD and Emily Johnson MD.

THE NBME CLINICAL NEUROLOGY SHELF EXAM

The neurology shelf exam has a reputation as one of the most challenging shelf exams. There are 110 questions and the test lasts 2 hours 45 minutes. The basic outline of the exam is available on www.nbme.org, shown here:

System
- General Principles, Including Normal Age-Related Findings and Care of the Well Patient — 1%–5%
- Behavioral Health — 3%–7%
- Nervous System & Special Senses — 60%–65%
  - Infectious, immunologic, and inflammatory disorders
  - Neoplasms (cerebral, spinal, and peripheral)
  - Cerebrovascular disease
  - Disorders related to the spine, spinal cord, and spinal nerve roots
  - Cranial and peripheral nerve disorders
  - Neurologic pain syndromes
  - Degenerative disorders/amnestic syndromes
  - Global cerebral dysfunction
  - Neuromuscular disorders
  - Movement disorders
  - Paroxysmal disorders
  - Sleep disorders
  - Traumatic and mechanical disorders and disorders of increased intracranial pressure
  - Congenital disorders
  - Adverse effects of drugs on the nervous system
  - Disorders of the eye and ear
- Musculoskeletal System — 10%–15%
- Other Systems, Including Multisystem Processes & Disorders — 15%–20%
- Social Sciences, Including Death and Dying and Palliative Care — 1%–5%

Physician Task
- Applying Foundational Science Concepts — 10%–15%
- Diagnosis: Knowledge Pertaining to History, Exam, Diagnostic Studies, & Patient Outcomes — 55%–60%
- Health Maintenance, Pharmacotherapy, Intervention & Management — 25%–30%

Site of Care
- Ambulatory — 60%–65%
- Emergency Department — 25%–30%
- Inpatient — 5%–15%

Patient Age
- Birth to 17 — 10%–15%
- 18 to 65 — 55%–65%
- 66 and older — 20%–25%

Neurology is a broad subject. At the most basic level you must be familiar with the anatomy and function of the brain and spinal cord, the most complex organs in the body. In addition you must be familiar with the brain's vascular supply to understand stroke, with immunology to understand multiple sclerosis, with infectious disease to treat and manage meningitis – the list goes on. Don't be intimidated by the large volume of information in the specialty – if you can master the neurologic exam, have a basic understanding of neuroanatomy and localization and know the principles of management for the most common illnesses outlined in this book, you will go far.

Most students in the neurology clerkship will pursue training in other medical or surgical specialties. No matter which field you go into, knowledge of neuroscience and neuroanatomy will be useful. Internists and family practitioners are often the first to manage headache, hear about TIA's or see early signs of ALS. Obstetricians must deal with young woman with multiple sclerosis, epilepsy and migraines during pregnancy. Anesthesiologists face the challenge of sedating patients with myasthenia gravis or muscular dystrophy. Surgeons must have knowledge of peripheral neuroanatomy to avoid inadvertent nerve injury.

It is my hope that this book will not only allow you to excel on the NBME clinical neurology shelf and USMLE Step 2 exam, but that it will provide you with the clinical knowledge and resources that you need for the rest of your career in medicine. This clinically oriented textbook will serve as an excellent resource during residency training and beyond.

Paul D. Johnson, MD

Neuroanatomy

"It's just a flesh wound."
Monty Python and the Holy Grail

In this section we will explore the basics of functional neuroanatomy and the correlates on neuroimaging. There is a tight connection between structure and function in the nervous system – injuries in discrete areas of the brain, spine or nerves reliably result in discrete symptoms. For example, an injury to the superior medial motor cortex of the brain will reliably cause weakness in the opposite lower extremity. As another example, you will come to see that a left frontal lobe injury will reliably cause both right sided weakness and language dysfunction, because the motor cortex and language centers are located very close to each other.

The nervous system can be divided into three parts – starting with the central nervous system (CNS), which includes the brain, brainstem, cerebellum and spinal cord. The peripheral nervous system (PNS) consists of the peripheral nerves, both motor and sensory, which connect the spinal to muscles and sensory organs throughout the body. Lastly, the autonomic nervous system, which is really a subset of the peripheral nervous system, controls bodily activities unconsciously, such as heart rate, blood pressure, sweating, and digestion.

Central Nervous System

The brain consisting of two cerebral hemispheres

The brainstem, connecting the brain to the spinal cord and cerebellum and housing many critical functions

The cerebellum, responsible for coordination and smooth movement

The spinal cord, comprising cervical, lumbar and thoracic segments and containing the descending motor neurons and ascending sensory neurons.

Motor Neurons originate in the motor cortex of the brain and travel down the spinal cord, where they synapse with peripheral nerves. Motor nerves in the central nervous system are called **upper motor neurons.**

Peripheral Nervous System

Nerve axon, the body of the nerve – this single cell can be several feet long.

Nerve ending

Nerve root attaches to spinal cord

Cell body

Muscle

Peripheral motor neurons are known as **lower motor neurons**. They synapse in the spinal cord and connect directly to muscle- when they are stimulated they cause muscles to contract.

Teaching Point: All neurons have an inner cell body, called an **axon**, as well as an outer layer called **myelin**. The myelin acts like the rubber insulation on electrical wires, and is critical for proper nerve function. There are differences between the myelin in the central and peripheral nervous systems, so some diseases will affect one and not the other.

Upper Motor Neuron lesions	vs	Lower Motor Neuron lesions
▪ From brain or spine injury		▪ Peripheral nerve injury
▪ Increased tone		▪ Decreased tone – flaccid
▪ Increased reflexes		▪ Decreased reflexes
▪ Minimal atrophy		▪ Significant atrophy

▲ *The pattern of atrophy, change in tone and change in reflexes can help localize injuries to the central or peripheral nervous system. All findings above are chronic.*

Grey matter refers to the collection of cell bodies. A group of cell bodies in the central nervous system is called a **nuclei**, whereas a group of cell bodies in the peripheral nervous system is called a **ganglion**.

White matter refers to the myelinated cell axons. Bundles of axons within the central nervous system are **tracts**, and within the peripheral nervous system they are called **nerves**.

The Lobes of the Brain

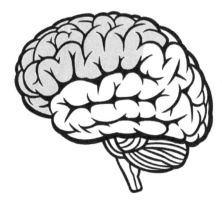

▲ *Frontal lobe – the anterior portion provides 'executive function' including planning, socially appropriate behavior, and the motor cortex is located more posteriorly.*

▲ *Parietal lobe - somatosensory cortex, which is required for sensation, identifying objects by touch, and attention.*

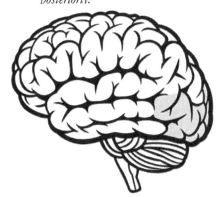

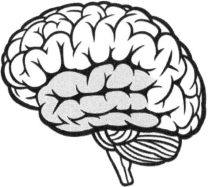

▲ *Occipital lobe - visual cortex, central vision localizes to the most posterior part of the occipital lobe.*

▲ *Temporal lobe - memory, sensory integration, auditory cortex. Contains the hippocampus which is necessary for memory formation.*

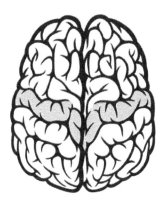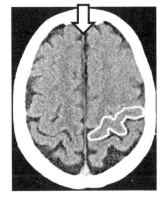

▲ *Motor cortex (outlined in white on head CT). The cell bodies of upper motor neurons are found here. Both the motor and sensory cortex extend down into the space between the two hemispheres, known as the interhemispheric fissure (arrow).*

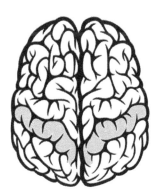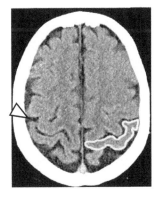

▲ *Somatosensory cortex (outlined in white on head CT), where peripheral sensory inputs are processed. Located just posterior to the motor cortex and separated from it by the central sulcus (arrowhead), which is the large sulcus that extends to midline and also serves to separate the frontal and parietal lobes.*

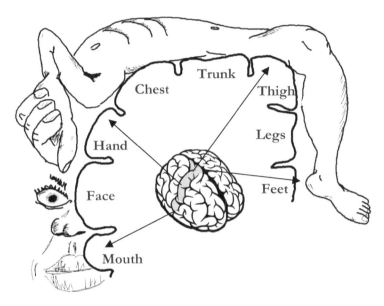

▲ The **Homunculus** *(Latin for "little man") showing how the body is represented within the brain. Both the motor and the sensory cortex share this anatomical layout. A relatively large amount of brain cortex is dedicated to the face and hand, given their relatively small size as compared to larger body parts such as the chest and trunk. This is due to the high density of sensory nerves in the hand and face. One clinical implication is that a single lesion will often affect both the hand and face, given their close proximity. This 'somatotopic' - or point for point anatomical relationship- is maintained in the descending nerve fibers and spinal cord as well. Note also that the leg motor and sensory cortex wraps around and down into the interhemispheric fissure.*

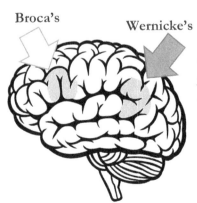

◄ *Language centers is in the dominant cerebral hemisphere – which in right handed people and most half of left handed people is the left hemisphere. The anterior location (white arrow, called Broca's area) largely controls speech production, while the posterior area (dark arrow, called Wernicke's area) is responsible for comprehension. The two cortical areas are linked internally.*

The Brainstem

The three components of the brainstem are the **midbrain, pons and medulla**. The brainstem connects to the thalamus above and the spinal cord below, as well as to the cerebellum. The thalamus is the 'relay station' for all sensory inputs into the brain, and is necessary to maintain consciousness. The fourth ventricle drains CSF from the lateral ventricles out to the spinal canal.

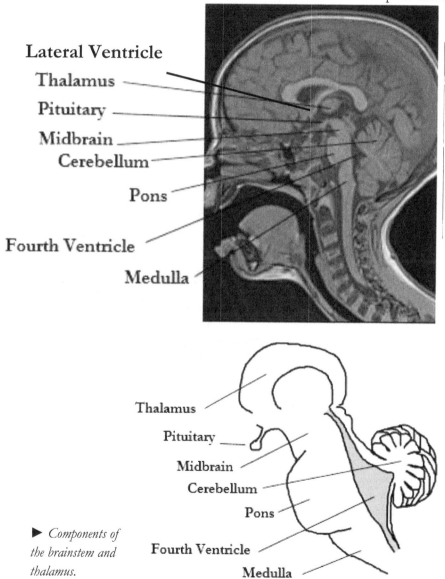

▶ *Components of the brainstem and thalamus.*

The Cranial Nerves

There are 12 paired cranial nerves (CN), each pair has a very specific, highly developed function. Cranial nerves three through twelve (often indicated with the roman numericals III-XII) arise directly from the brainstem. Cranial nerves I and II, the olfactory and optic nerves respectively, arise directly from the brain itself.

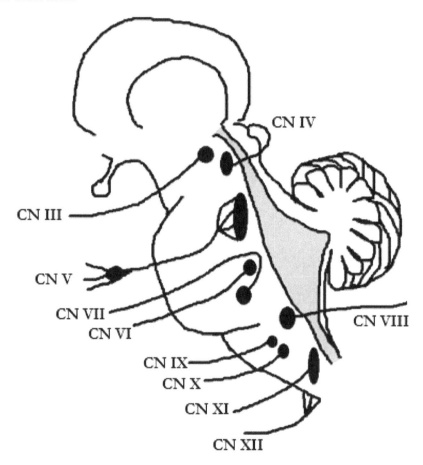

▲ *Cranial nerve diagram. Note that CN III-IV arise from the midbrain, V-VIII from the pons and IX-XII from the medulla. Cranial nerve IV is the only one that arises from the back of the brainstem, and loops around to the front. CN VII wraps around CN VI. CN V, the trigeminal nerve, forms a small nucleus outside of the brainstem before giving off three distinct sensory branches.*

The Cranial Nerves

Cranial Nerve	Function	Clinical Pearls
I Olfactory N.	Provides the sense of smell	Rarely tested, lost early in Parkinsons
II Optic N.	Transmits vision from the eyes to the occipital cortex	The optic disc, seen on the funduscopic exam, is the head of the optic nerve
III Oculomotor N.	1) Moves the eye in 4 of the 6 cardinal directions (see diagram) 2) Constricts the pupil	Diabetes can impair CN III motor function, leading to an eye that is "down and out." Compression from an aneurysm or herniation may cause a dilated pupil
IV Trochlear N.	Tilts the eyes down and in (so you can look at the tip of your nose)	Dysfunction can cause a subtle vertical diplopia, patients compensate with a head tilt to the opposite side
V Trigeminal N.	Provides sensation to the face, including the cornea, and innervates the muscles of the jaw	Often implicated in sensory changes in the face
VI Abducens N.	Allows the eye to abduct (look laterally)	Susceptible to injury from increased intracranial pressure, Wernickes & trauma
VII Facial N.	Motor function of the face, and taste to the anterior 2/3 of the tongue	CNS injuries typically affect just the lower face, whereas a peripheral CN VII injury causes upper and lower facial weakness
VIII Vestibular N.	Hearing and balance	Viral inflammation leads to vertigo

IX Glossopharyngeal N.	Swallowing, control of the soft palate	Test by checking gag Uvula deviates away from the lesion
X Vagus N.	Provides extensive autonomic control of the heart, lungs and digestive tract, as well as swallow function	Stimulation of the vagal nerve slows the heart rate - over compensation during stress can lead to syncope
XI Spinal Accessory N.	Controls shoulder shrug and head turning	Injury to this nerve is most often due to surgery or trauma
XII Hypoglossal N.	Provides motor control of the tongue	Tongue deviates towards the lesion

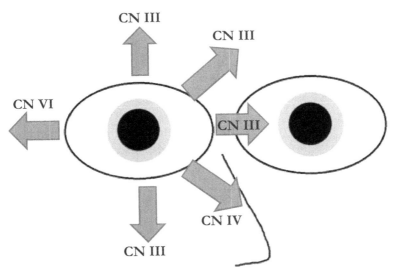

▲ *Diagram of the cranial nerve control of the eye. Notice that CN III controls four of the 6 directions of movement. CN VI only abducts, and CN IV only looks at the tip of the nose. When CN III is injured the eye tends to drift out laterally, because it is pulled by CN VI, and down, because it is pulled by CN IV.*

◄ *The 'down and out' eye found in a CN III palsy. There is often severe ptosis as the 3rd nerve helps to keep the eyelid up.*

Visual Fields

Monocular vision loss is almost always due to a problem in the eye or in the optic nerve. The visual input from each eye merges in the optic chiasm, and is separated into left and right hemifields in the optic tracts. This means that each optic tract contains one-half of the visual input from each eye – the left visual fields go through the right optic tract and right hemisphere, and the right visual fields pass through the left optic tract and therefore the left hemisphere. The optic radiations also separate into superior and inferior visual fields – nerves with superior visual field information run through the temporal lobes. Inferior visual field information is found in the parietal lobes. Each ends up on the banks of the Calcarine fissure in the occipital lobe.

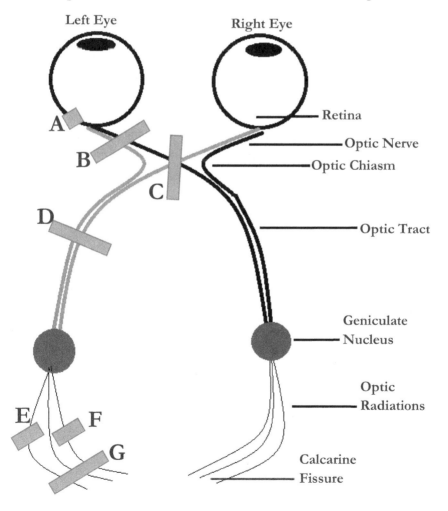

Visual Field Defects

This page is high yield for the NBME shelf exam.

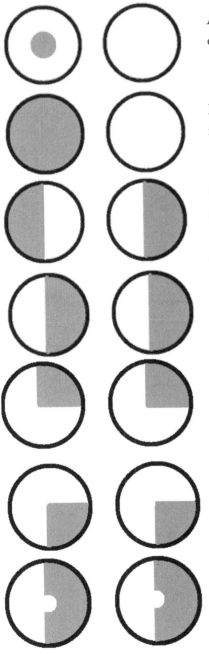

A) Retinal injury resulting in central scotoma.

B) Optic nerve injury with monocular vision loss.

C) Bitemporal hemianopsia; crossing temporal visual field fibers from each eye are affected in an optic chiasm injury.

D) Optic tract injury with contralateral homonymous hemianopia.

E) Temporal lobe optic radiation injury, contralateral superior quadrantopia.

F) Parietal lobe optic radiation injury, contralateral inferior quadrantopia.

G) Contralateral homonymous hemianopia – may or may not have macular sparing.

Frontal lobes are responsible for executive function, planning and self-control.

Parietal lobes process sensory input, such as touch, pain, proprioception and attention. Injury can cause patients to 'neglect' their contralateral side.

Occipital lobes process vision – remember the occipital lobes are both flipped and inverted from actual vision, so a lower right occipital lobe injury causes an upper left visual field loss!

Temporal lobes contain the hippocampus, which is responsible for forming new memories. Temporal lobe injuries create a high risk for epilepsy.

Cerebellum responsible for coordination of fine movements – injuries here can cause vertigo, ataxia or clumsiness.

Corpus callosum the white matter structure that connects the right and left hemispheres. It is rarely affected by stroke, but commonly involved in inflammatory conditions such as multiple sclerosis.

Putamen, Globus Pallidus and Caudate together form the basal ganglia, responsible for motor functions and posture among other things, and is especially important in movement disorders like Parkinson disease.

Internal capsule represents the descending motor and sensory nerves. This location is vulnerable to lacunar strokes, which are small, 1.5 cm or less.

Thalamus maintains wakefulness, but is also a 'relay station' for all ascending sensory and motor fibers. A very important brain structure – injury here can cause paralysis, loss of sensation or lethargy and somnolence.

Lateral and fourth ventricles are normal fluid filled spaces within the brain. Cerebrospinal fluid (CSF) is produced within the lateral ventricle, flows out through the fourth ventricle to surround the spine, and is then reabsorbed. Blocking the flow of CSF can lead to hydrocephalus.

Sylvian fissure separates the frontal lobe from the temporal lobe. The middle cerebral artery runs through this space.

Falx cerebri and tentorium cerebelli are extensions of the meninges, acting as support structures for the brain. In significant brain injuries swelling can cause brain tissue to move (herniate) from one side of these structures to another – an ominous sign!

Descending Motor Pathways

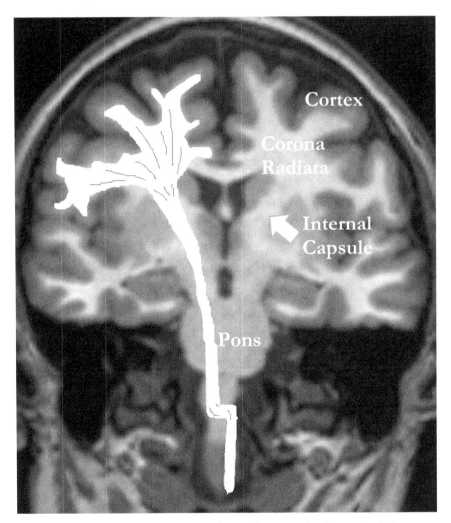

▲ *Diagram of the descending motor pathways, known as the Corticospinal Tract, shown in white. The motor neurons arise from the motor cortex, and nerve axons funnel down through the corona radiate into the internal capsule, where they are tightly bundled, and move down through the brainstem, including the midbrain, the pons – where facial motor fibers break off to form CN VII, and down through the medulla. Note that at the bottom of the medulla the motor fibers* **cross to the opposite side**, *or decussate (*). This crossing-over is why each cerebral hemisphere controls the* **opposite side** *of the body (i.e. why a right cortical stroke causes left sided weakness).*

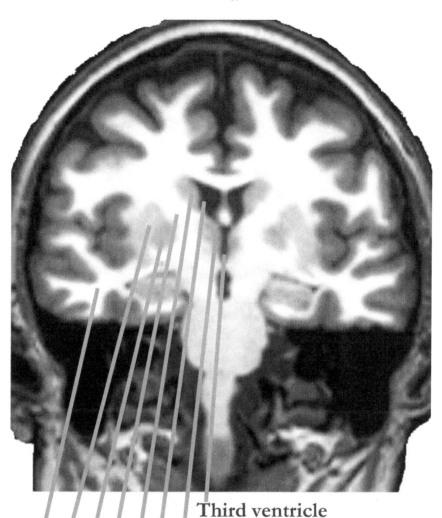

▲ *Coronal T1W brain MRI*

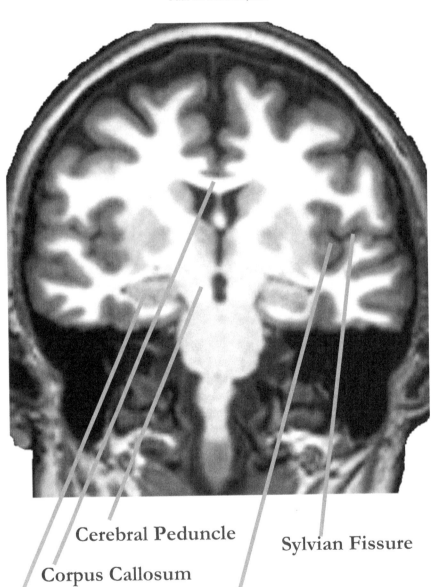

▲ *Coronal T1W brain MRI*

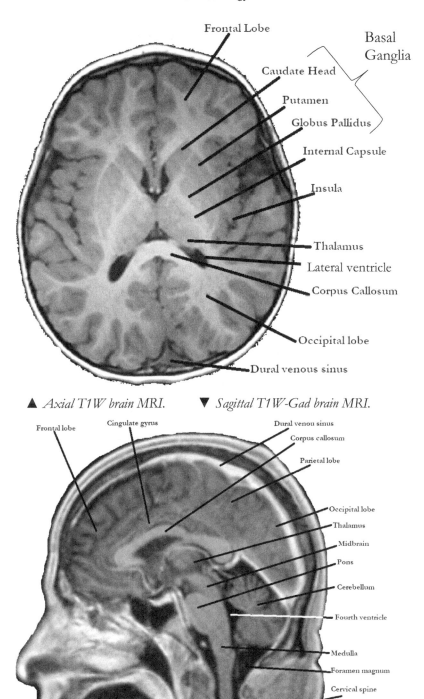

▲ *Axial T1W brain MRI.* ▼ *Sagittal T1W-Gad brain MRI.*

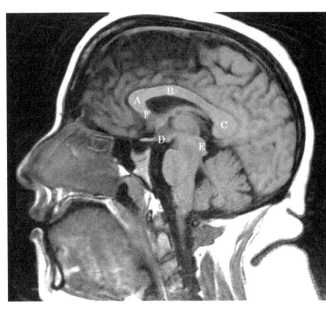

A - Genu of the corpus callosum

B - Body of the corpus callosum

C - Splenium of the corpus callosum

D - Mamillary bodies

E - Tectum of the midbrain

F - Rostrum of the corpus callosum

▲ *T1W Sagittal brain MRI. The bright white corpus callosum is a tip off that it is T1W image. The corpus callosum is the primary connection between the two hemispheres – although there are a few, smaller, connections as well.*

Tip: Keep in mind that motor and sensory fibers cross in the brainstem, so that right body motor and sensory are actually controlled by the left brain. The visual fields not only cross sides, but they flip vertically as well – so, for example, right upper visual fields are processed in the lower left part of the occipital lobe. One structure that isn't crossed is the cerebellum – a cerebellar lesion will cause ataxia on the same side as the injury.

The Extracranial Circulation

These are the blood vessels connecting the heart to the brain. Two carotid arteries and two vertebral arteries arise from the aortic arch and the subclavian arteries. The two carotids form the anterior circulation, supplying blood to the frontal, parietal and lateral temporal lobes. The two, smaller vertebral arteries join at the base of the brain to form a single basilar artery, which supplies the brainstem, occipital lobes and medial temporal lobes.

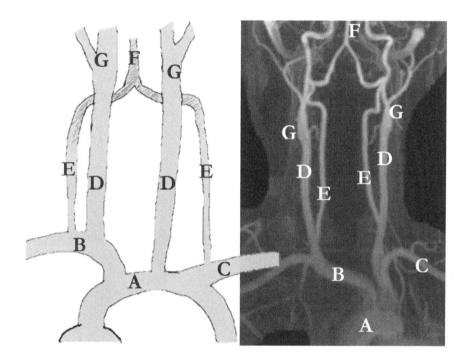

▲ *Diagram of the extracranial circulation (left) and MR angiogram (right) of the same vessels. A) aortic arch, B) right subclavian artery, C) left subclavian artery, D) common carotid artery, E) vertebral artery, F) basilar artery and G) carotid bifurcation – where the common carotid splits into the internal carotid and external carotid arteries. The internal branch supplies the eyes and brain, while the external carotid supplies the soft tissue of the face and scalp.*

Looking for more neuroanatomy resources? I recommend Stephen Goldberg's *Clinical Neuroanatomy Made Ridiculously Simple* and the website www.headneckbrainspine.com

The Intracranial Circulation

Intracranial circulation is divided into the anterior circulation, which consists of the internal carotid arteries and their branches- the middle and anterior cerebral arteries. The posterior circulation consists of the basilar artery, which supplies the brainstem, and it's branches- the posterior cerebral arteries, which supply the thalamus, occipital lobes and medial temporal lobes.

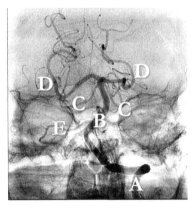

◄ *Posterior Circulation* Cerebral angiogram, vertebral injection and coronal view, showing A) the left vertebral artery, B) the basilar artery, C) superior cerebellar artery, D) posterior cerebral arteries (PCA), and E) posterior inferior cerebellar artery (PICA).

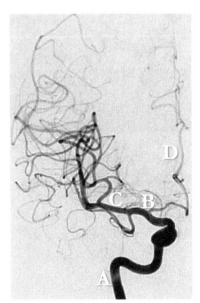

◄ *Anterior Circulation* Cerebral angiogram, carotid injection and slight lateral view, showing A) the internal carotid artery, B) carotid bifurcation, C) middle cerebral artery (MCA) and, D) the anterior cerebral artery (ACA).

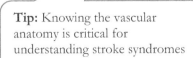

Tip: Knowing the vascular anatomy is critical for understanding stroke syndromes

The Circle of Willis

Although we've shown the anterior (carotid) and posterior (vertebrobasilar) circulations separately, there are actually a number of small connections between the two, which provides important collateral circulation. The 'Circle of Willis' is the network of small anastamotic branches – consisting of the anterior communicating connecting the two anterior cerebral arteries, and the posterior communicating, which links the internal carotids to the posterior cerebral arteries.

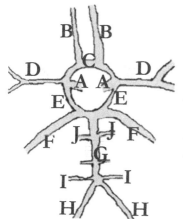

◀ *Circle of Willis Diagram showing the A) internal carotid arteries, B) anterior cerebral arteries, C) anterior communicating artery, D) middle cerebral arteries, E) posterior communicating arteries, F) posterior cerebral arteries, G) basilar artery with perforating branches, H) vertebral arteries, I) posterior inferior cerebellar arteries, and J) superior cerebellar arteries..*

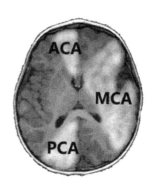

▶ *Major vascular territories within the brain. The Anterior Cerebral Artery (ACA) supplies the midline frontal lobes. The Middle Cerebral Artery (MCA) supplies frontal and parietal lobes, and supplies the largest brain area. The Posterior Cerebral Artery (PCA) supplies the occipital lobes, medial temporal lobes, and thalamus.*

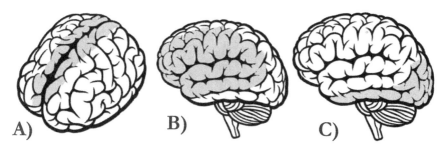

▲ *A) anterior cerebral artery territory, B) middle cerebral artery territory and C) posterior cerebral artery territory.*

The Spinal Cord connects the brain to the peripheral nervous system. It is a compact bundle of descending motor neurons and ascending sensory neurons. Some reflexes (i.e. withdrawal) may be directly controlled by the spinal cord, facilitating quick reaction times.

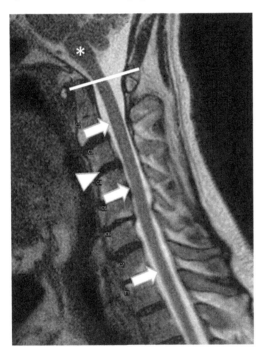

◀ T2W *cervical spine MRI showing the medulla (*) and spinal cord (arrows). The foramen magnum (white line) is the opening at the bottom of the skull, connecting to the spinal canal. Note the intervertebral discs (arrowhead). You should be able to see spinal fluid surrounding the cord. Note that spine MRI is technically challenging and studies are often poorer quality than brain MRI scans.*

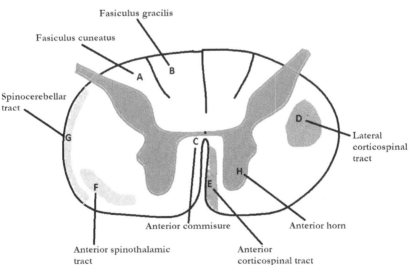

▲ *Spinal cord anatomy (cross-section of cervical spinal cord).*

Components of the spinal cord include the **lateral corticospinal and anterior corticospinal** tracts, which carry descending motor information from the contralateral cerebral hemisphere. The **dorsal columns** (fasciculus gracilis and cuneatus) carry ascending vibration and proprioception information to the brain. The **spinothalamic** tract carries ascending pain and temperature sensation. Spinothalamic fibers are unique in that they cross to the contralateral side almost immediately within the spinal cord before ascending, rather than crossing in the medulla as the dorsal columns do – this means that an injury to one half of the spinal cord could cause a loss of pain and temperature sensation on one half of the body, and loss of vibration/proprioception and weakness on the other half – known as **Brown-Séquard** syndrome.

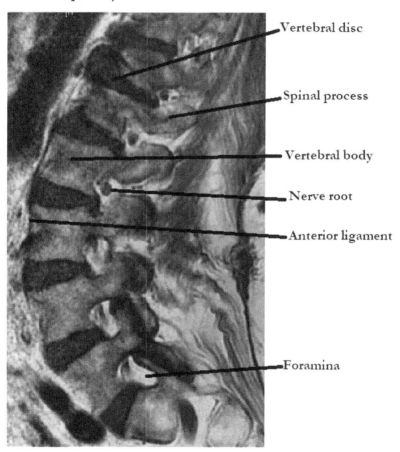

▲ *T2W para-sagittal lumbar spine MRI. You can imagine how a herniated intervertebral disc could impinge on the nerve root as it exits the foramina, causing a radiculopathy (nerve root injury).*

The Lumbar Puncture

A lumbar puncture can generally be performed safely at the bedside, as long as you can palpate good landmarks. Instructions on how to perform a lumbar puncture safely are outside of the scope of this book, but we will cover some basics of CSF analysis here. Because the spinal cord ends at approximately L1-L2 in most people, the procedure should be done below L2-L3. Opening pressure should be routinely check in the lateral decubitus position (measurements are not accurate if the patient is sitting up).

> **Key CSF Lab Tests – send in all patients**
>
> - ✓ Protein
> - ✓ Glucose
> - ✓ Cell count and differential
> - ✓ Gram stain and culture
>
> **Send if you suspect Multiple Sclerosis**
>
> - ✓ IgG Index (requires a concurrent blood draw)
> - ✓ Oligoclonal bands

Contraindications to Lumbar Puncture

- Platelet count < 50,000/mm^3
- INR > 1.4
- Infection over the site of the lumbar puncture
- Intra-cranial mass lesion (no specific criteria for size exists)
- In general, anticoagulants should be held for 5 half-lives (at least 4-6 hours for a heparin infusion)

Complications of Lumbar Puncture

- Orthostatic headache is most common, and typically related to low intra-cranial CSF volume
- Infection
- Bleeding
- Nerve damage (rare)

> **Tip:** A blood patch can be curative in cases of prolonged post-lumbar puncture headache, if fluids, time and drinking caffeine don't help first.

Normal CSF Lab Values

- ✓ **Protein:** 15-45 mg/dL, higher after surgery, and increases with age and history of diabetes
- ✓ **Glucose:** 45-80 mg/dL, or 60% of the serum glucose; very low glucose levels suggest bacterial or fungal infection
- ✓ **Cell count:** should not exceed 5 WBC's/mm³, no RBC's; pleocytosis (excess white cells in the CSF) suggests infection or inflammation, but can also be seen transiently after a seizure
- ✓ If the CSF is contaminated by hemorrhage, correct for 1 extra WBC per extra 750 RBC's

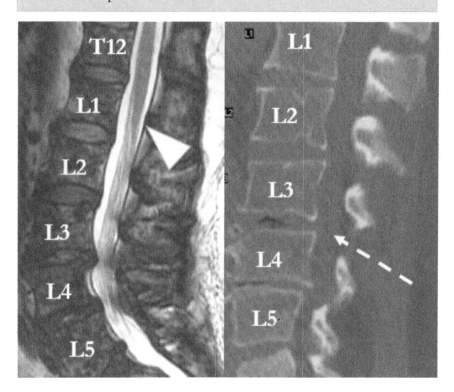

▲ *T2W sagittal spine MRI on the left. The spinal cord generally ends at L1 (arrowhead), the bottom of the cord is known as the conus medullaris. Below this hang descending lumbar and sacral nerves, called the cauda equina. On the spine CT on the right, notice how the CT scan demonstrates the vertebral bones well, but does not show the spinal cord. The dashed line demonstrates the trajectory for a lumbar puncture, aiming for the L3-4 interspace, well below the level of the spinal cord.*

Notes

Exam Pearls

"We have top men working on it right now."
- "Who?"
- "Top men."
Raiders of the Lost Ark

Pearls for the Neurologic Clinical Exam

The neurologic exam is a way of assessing the function of the brain, spine and nerves, and is incredibly useful in identifying and localizing neurologic disease. The purpose of this section is to highlight some of the key aspects of the examination, and my personal approach.

Handedness

As a general rule of thumb in neurology it is helpful to know the patient's handedness, i.e. right handed, left handed or ambidextrous. Handedness provides insight into which cerebral hemisphere is dominant, which is helpful because language localizes to the dominant hemisphere. Most people are right handed and left hemisphere dominant. A portion of people who are left handed will also be left hemisphere dominant. Knowing handedness can help you to make sense of confusing situations when language deficits coincide with apparent right hemisphere injuries, and may soon be important for billing.

Basic Neurological Exam Components

Note that a complete and billable neurology note requires *at least* two components of the cardiovascular exam <u>and</u> a funduscopic exam.

Mental Status
- Orientation
- Language

Cranial Nerves
- Gaze & pupillary function
- Visual fields
- Facial symmetry
- Dysarthria

Motor
- Bulk, tone and strength

Sensation
- Neglect, Agnosia
- Romberg

Coordination
- Finger-to-nose and heel-to-shin

Reflexes

Gait

The mental status exam

I will often ask a few screening questions for the mental status exam, and obtain a detailed mental status examination if there is any concern for change in thinking, altered behavior or abnormal interaction. Good screening questions include asking about the date and current location, asking the patient their age and performing a three item recall test. For three item recall I typically use the same items each time – mine are "apple, penny, table." Have the patient repeat the items at least once to ensure that they have heard them appropriately, and then ask them recall the same items in 3-5 minutes.

The components of the Mini-Mental Status Exam (MMSE) © are given here, or see page 216 for more details, including a description of the Montreal Cognitive Assessment (MOCA) tool ©. Significant abnormalities on any of these tests should prompt referral for formal neuropsychological testing.

Mini-Mental Status Exam

Orientation
- What is the day, date, month, season and year? *(5 pts)*
- Where are we? Country, state, city, hospital and floor? *(5 pts)*

Registration
- Name 3 objects, make sure patient can repeat all 3 *(3 pts)*

Attention & Calculation
- Serial 7's, stop after 5 answers *or* spell WORLD backwards *(5 pts)*

Recall
- Name the 3 objects repeated above *(3 pts)*

Language
- Name a pencil and a watch *(2 pts)*
- Repeat "No ifs, ands, or buts" *(1 point)*
- Follow a 3 step command *(3 pts)*
- Copy a simple design *(1 point)*
- Read and obey the command: "Close your eyes" *(1 point)*
- Patient writes a sentence of their choice *(1 point)*

Total: 30 points possible

▲ *Normal is 24 points or higher, 19-23 points represents mild cognitive impairment.*

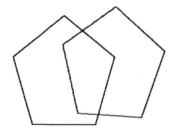

◀ *In the Mini-Mental Status Exam (MMSE) patients are asked to copy a simple design, which is typically a set of interlocking pentagons. Note that the MMSE is copyrighted and shouldn't be used in research without permission.*

Testing language

Language function occupies a large portion of the dominant cerebral hemisphere, and abnormalities with speech and language are common neurological complaints. *Dysarthria* is an abnormality with speech production or phonation, usually due to dysfunction of the pharynx, tongue or lips. *Aphasia* is a disorder of language, including grammar, syntax and vocabulary. The two are frequently confused by non-neurologists. The non-fluent aphasias, characterized by limited speech production, tend to be due to lesions more anterior in the brain, near the frontal motor cortex. So called fluent aphasia, characterized by rapid, fluent nonsensical speech, is typically due to a more posterior brain injury, for example a temporal lobe injury.

Components of the Language Exam

Fluency
- Normal speech includes > 8 word sentences, speaking 100-115 words per minute

Comprehension
- Start with simple commands and build up to more complex tasks, i.e. *touch your left ear with your right thumb* is complex
- Note – imitation is <u>not</u> comprehension

Naming
- Ask patients to name several objects, both common and uncommon (ex. watch, wristband and watch clasp)

Repetition
- Start with simple sentences and increase complexity

Reading/Writing
- In aphasia reading/writing are also affected

▶ *Diagram of the brain, showing the anterior (*) and (★) posterior language centers (connected subcortically). Recall that anterior lesions tend to cause non-fluent aphasia, while posterior lesions result in fluent aphasias.*

The Aphasia Flow-Chart

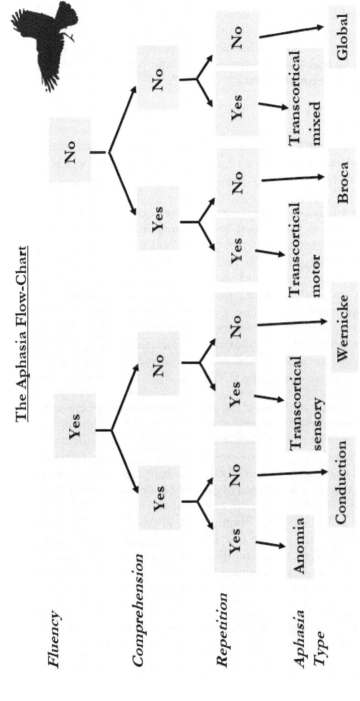

▲ *If you assess fluency, comprehension, repetition and naming, you will be able to categorize aphasias by type.*

Cranial nerve exam

Refer to pages 8-10 in the anatomy section for the location, name and function of each cranial nerve. Significant time can be spent on the cranial nerve exam when brainstem or basilar artery disease is suspected. In general, I like to start by checking pupillary response and palate elevation, because both of these tests require a light.

Optic nerve (CN II)

When checking pupillary response you are looking for pupillary asymmetry, which is called *anisocoria*. Up to a 1 mm difference between pupils is normal – refer to the text box *Evaluating Anisocoria* on the next page for more details.

Normally, the pupils will dilate or constrict based on the amount of ambient light, constricting in bright light and dilating in dim light to maintain adequate vision. Light hits the eye and retina and travels through the optic nerve to the midbrain, where light intensity is registered. A signal is then sent back to the eye through the third cranial nerve, causing the pupil to adjust to the amount of ambient light. *Unilateral* dysfunction of the eye/retina/optic nerve can cause a relative afferent pupillary defect, or rAPD, in which the pupils fail to respond appropriately when light is shined in one eye at a time. This is important because inflammation of the optic nerve, a very common symptom in multiple sclerosis and other demyelinating diseases, can cause an rAPD.

To detect an rAPD you must shine a bright light into one eye at a time. When you shine a bright light into a normal eye with a healthy optic nerve the brain 'sees' bright ambient light and responds by constricting pupils. When you swing the flashlight over to an eye with an inflamed and dysfunctional optic nerve, not as much of the light signal will get through – even though the intensity of your flashlight hasn't changed, the brain will 'see' less light – it will seem to the brain that the room has gotten darker. The response will be dilation of both pupils. In the swinging flashlight test you move a flashlight from eye to eye and see pupillary constriction when the light is in the normal eye and dilation when the light is shined in the affected eye. Be aware that anything that blocks light transmission in one eye, such as bad cataracts or retinal disease, can also cause an rAPD.

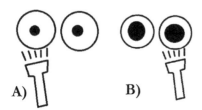

◄ *Example of a relative afferent pupillary defect. Here B) shows the light shining into the abnormal eye. See page 234 for a clinical example.*

Evaluating Anisocoria

When pupillary asymmetry is present the first step is deciding which is the abnormal side – the larger or smaller pupil? Examining the patient in bright and dim light will help to clarify, as the abnormal pupil will be poorly responsive to the change in light.

The Blown Pupil
- May be a sign of 3rd nerve palsy, usually due to compression
- Compression can be cause by an anterior cerebral artery or posterior communicating artery aneurysm, which can be evaluated for by CT angiogram
- Medications are a *very* common cause of unilateral anisocoria in the hospital – for example, a drop of albuterol in one eye will cause that pupil to dilate

The Constricted Pupil
- As in dilated pupils, consider medications such as opiates, antipsychotics, or cholinergic medicines
- Can also be due to injuries in the sympathetic pathway, which runs down the brainstem, into the upper thoracic spinal cord, over the apex of the lungs, and up the carotid artery and back to the eye (whew!)
- Therefore, a carotid dissection could cause unilateral pupillary constriction (miosis), as well as subtle ptosis

Extra-ocular movements (CN III, IV and VI)

In normal vision the gaze is *conjugate*, meaning both eyes are aligned and move together equally. Sedated, comatose or asleep patients will develop a disconjugate gaze, where each eye drifts out laterally, which is the natural position for the eyes when the extra-ocular muscles are completely relaxed.

Take note of whether gaze is smooth or choppy – loss of *smooth pursuit* can indicate a cerebellar dysfunction (recall, the cerebellum helps 'smooth out' movements).

Observe for nystagmus, in which there is rapid, small amplitude, beating movement of the eyes. The eyes will drift slowly to one side and then jerk back quickly to midline. The direction of nystagmus is named for the direction of the fast phase. Nystagmus can be normal, pathologic, congenital or even voluntary. It can be due to drug toxicity (especially alcohol, sedatives, and many anti-seizure medications such as phenytoin). It can be due to peripheral vestibular disease, such as benign paroxysmal positional vertigo (BPPV) or due to central disease such as a cerebellar stroke. Physiologic nystagmus occurs with extreme lateral gaze, in which several beats of nystagmus in the direction of gaze are normal. **Red flags for nystagmus** include:

> ➢ Alternating directions of nystagmus (i.e. beating strongly to both the right and left sides)
> ➢ Vertical nystagmus (upbeat or downbeat)
> ➢ Prolonged vertigo and nystagmus, without any triggers or without improvement over time

Benign Paroxysmal Positional Vertigo

BPPV is characterized by sudden, severe onset of vertigo and nystagmus, occasionally so severe that it causes vomiting. BPPV is triggered by head movements – the symptoms should be brief, lasting minutes or less, and improve significantly with rest. It is a peripheral disease caused by debris within the semicircular canals, which disrupts the patient's sense of equilibrium. BPPV can be diagnosed at the bedside with the classic history and the **Dix-Hallpike** maneuver. It can be treated with the **Epley** maneuver. Demonstrations of both of these are readily available on the internet. Keep in mind that head movement will worsen vertigo of *any cause*, but in BPPV symptoms should *only* be provoked by specific head movements. BPPV is especially common after minor head trauma, which can cause the inner ear debris to break off in the first place. BPPV can occur in children.

Visual fields

Although the sensitivity of checking visual fields at the bedside is relatively poor, it is an important test, as the occipital lobes represent a large portion of the brain and pathology in that area can be missed if visual fields are not checked. Confrontational visual fields must be tested in each eye individually – when both eyes are open there is significant overlap in vision, and visual field defects can be missed.

Have the patient cover one eye at a time. Stand arm's length away - about 3 feet. You can close your corresponding eye, if you like, to get a sense of what the patient's visual field should be. Ask the patient to look at your nose, and give them a visual stimulus in each of the four quadrants of view. You can flash a finger or ask them to count fingers. Cooperation is critical. If you suspect severe visual field defects you can always refer patients to ophthalmology for formal visual field testing. Unfortunately, even the best formal visual field testing by perimetry is reliant on a cooperative patient.

An easy bedside test for hemianopsia (loss of half of the visual field in both eyes) is to draw a horizontal line and then ask the patient to draw a vertical line exactly in the middle. If they have hemianopsia, they will draw their vertical line in the middle of the page *as they see it,* which will not be midline.

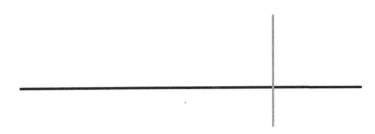

▲ *A patient with left sided hemianopsia was asked to draw a vertical line in the middle of the horizontal line. Since they don't see on the left side, their 'middle' was skewed far to the right.*

The facial nerve (CN VII)

Facial weakness is a common neurologic complaint – recall the homunculus and the large spatial representation of the face in the brain (page 6). Facial weakness with ipsilateral arm and leg weakness is easy to localize to the brain, but isolated facial weakness can be more challenging.

The most important part of evaluating isolated facial weakness is determining if there is both upper and lower face involvement, or only lower facial involvement. An injury to the facial nerve itself will cause weakness of half of the face, both upper and lower. Note that occasionally involvement of the upper face is slightly delayed in a peripheral injury.

If the facial weakness is due to a brain injury, then usually only the lower half of face is involved and the forehead is spared. This is because the forehead receives *dual innervation,* meaning it receives input from both sides of the brain, whereas the lower face receives input from only one side of the brain. This quirk means that spared forehead strength, tested by eyebrow raise and ability to wrinkle the forehead, usually indicates a central cause of facial weakness.

◄ *Left peripheral facial palsy, A) at rest and B) smiling. Note the loss of forehead wrinkle, weak eyelid causing a seemingly 'more open' left eye, and lack of ability to fully open the mouth on the right or raise cheeks to smile. Photo used with patient permission.*

Idiopathic facial nerve weakness is called a Bell's Palsy (note, it is not called Bell's Palsy if a reason for the dysfunction is found). Bell's palsy is relatively common, and is often due to non-specific viral inflammation. It may develop slowly over one or two days, and usually improves spontaneously after several weeks or months. Treatment is steroids with or without antiviral medication. Saline eye drops and taping the eye at night may help prevent corneal injuries. Brain imaging is not mandatory for patients with no red flags, i.e. no history of cancer, immunosuppression, first episode of facial nerve palsy, age $\leq$ 55 years, and no other findings.

Dysarthria (CN VII, IX, X and XII)

Dysarthria is an articulation or phonation problem, not a disorder of language, which involves grammar, syntax and vocabulary. A common pitfall is that edentulous patients may have dysarthria, which should not be confused with dysarthria from a neurologic cause. If a patient presents with isolated dysarthria, cranial nerve dysfunction such as the gag reflex (CN X), shoulder shrug (CN XI), tongue function including evidence of tongue atrophy (CN XII) and facial nerve function (CN VII) should be tested.

Asking patients to repeat "puh-kuh-tuh" can be helpful to localize the source of dysarthria in difficult cases. The 'puh' sound is made by the lips, 'kuh' by the pharynx and 'tuh' requires the tongue. Phrases such as *Methodist Episcopal* which vary consonants and vowels can be helpful. Presence of a normal cough indicates intact vocal cord function.

A relatively common cause of nasal dysarthria is myasthenia gravis, a condition which is covered in the neuromuscular disorder section of this book. Motor neuron diseases such as ALS are uncommon, but can also cause dysarthria.

The motor exam

Begin the motor exam by assessing bulk and tone. When checking bulk you are assessing for muscle atrophy, which is a sign of lower motor nerve injury (i.e. peripheral nerve injury). Especially important is asymmetric atrophy, such as atrophy in one limb, or atrophy in one hand or in the tongue.

Tone is the resistance to passive movement across a joint. Test tone by passively moving the extremities while the patient is relaxed. Increased tone can be voluntary, as in someone who is not fully relaxed, or in demented patients who actively move with you despite your instructions (called paratonia). Conditions such as Parkinson disease will cause diffuse increase in tone across an entire range of motion – combined with tremor this causes a *cogwheel rigidity*. Upper motor neuron injuries, such as in stroke, result in spasticity - a speed dependent increase in tone. When you move a *spastic* joint slowly you feel little change in tone, but rapid movements will cause the joint to quickly 'lock-up.'

When looking for weakness due to a brain injury, you are looking for weakness in the corticospinal tract (page 11). The corticospinal tract is the primary mode by which we perform voluntary, complex motor movements and extends from the motor cortex, through the internal capsule, brainstem and spine. There are other motor pathways, that help maintain balance and posture automatically, which we won't address here.

The corticospinal tract preferentially innervates the extensor muscles in the upper extremities and the flexor muscles in the lower extremities. Subtle signs of corticospinal weakness would therefore include weakness of finger or wrist extensors. This leads to the typical finding in stroke patients, where the upper extremity is flexed and internally rotated and the lower extremity is extended. This occurs because the upper extremity flexor tone and lower extremity extensor tone are relatively spared and unopposed.

Exam techniques include a test for pronator drift, in which patients are asked to hold both arms outstretched with palms up and eyes closed. Pronation, or subtle inversion of the hand, is a sign of weakness. Orbiting, where patients are asked to rotate their hands or fingers around each other, is another sensitive test of weakness. When positive the weak limb stays relatively immobile while the stronger limb 'orbits' around it.

MRC Muscle Scale

0 – No contraction
1 – Flicker/trace movement
2 – Movement without gravity
3 – Movement against gravity
4 – Movement against gravity and resistance
5 – Normal power

◀ *Pronator drift of the patient's right arm. Note that the arm doesn't just drop slightly, more importantly it is internally rotated (pronated).*

The sensory exam

Sensation is the most subjective part of the neurologic exam. Pearls for the sensory exam include differentiating between 'positive' findings such as paresthesias, pain, or feeling like you have ants on your skin (*formication*) vs 'negative' findings such as loss of sensation or loss of position sense. Positive findings are more likely to be related to an excitatory condition, such as migraine headache, seizure or cerebral amyloid angiopathy, whereas stroke, which causes loss of function, will almost always result in negative symptoms such as loss of sensation.

A nice way to screen for normal sensation is to place a small object such as a quarter, paper clip or key, into the patient's outstretched hands with their eyes closed. This can be done at the end of the pronator drift exam. If they can identify the object with their eyes closed you know that peripheral sensation is intact, and have also tested stereognosis (the ability to identify an object by touch), a function of the parietal lobe.

Another sensory examination is the Romberg test. Patients are asked to stand with their feet together, first with eyes open and then with eyes closed. The patient is reliant on their sense of balance through the inner ear and the sense of proprioception, which is a function of the dorsal columns of the spinal cord. If the patient cannot stand in place for 30 seconds with their eyes closed, this may indicate a sensory/proprioceptive disorder.

Neglect is a unilateral loss of attention. Test for neglect by providing either a visual or a sensory stimulus to the patient on both sides of the body, first on one side at a time, then simultaneously. You <u>must</u> make sure the patient doesn't have unilateral sensory loss. When you touch both arms, both legs or both sides of the face very gently with the patient's eyes closed – the patient should register being touched in two places. If they don't, they may have neglect, which reflects contralateral parietal lobe dysfunction leading to the inability to attend to one side of space. Neglect can be complete, or partial, depending on the amount of stimulus required to overcome it. Suspect neglect with parietal lobe injuries and test with double simultaneous stimulation to identify subtle dysfunction.

> **Tip:** Patients with severe neglect may not pay attention at all to the affected side. These patients may not attend to an examiner speaking from the neglected side. Neglectful patients will commonly eat only half the food on their plate – not being aware that the other half exists. Left sided neglect is the most common, and is commonly due to dysfunction of the right parietal lobe.

The Hand Motor and Sensory Exam

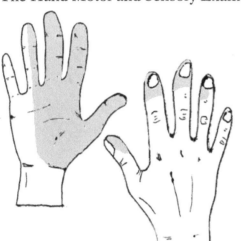

◀ **Median Nerve** *sensory innervation (grey). The most common location for median nerve compression is in the wrist (carpal tunnel syndrome), in which case the area of palmer sensory loss will be smaller.*

▶ **Median Nerve** *motor exam, the patient abducts the thumb – as if trying to make a 45 degree angle with the thumb and fingers. The muscle belly is the abductor pollicus brevis (APB, arrow).*

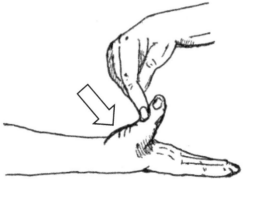

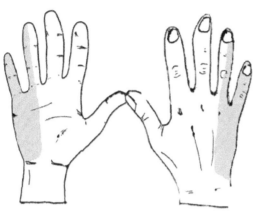

◀ **Ulnar Nerve** *sensory innervation (grey). The most common location for ulnar nerve compression is in the elbow (think of the 'funny bone').*

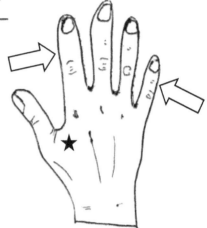

▶ **Ulnar Nerve** *motor exam, the ulnar n. innervates the intrinsic hand muscles. Test by pushing in against the pointer and little fingers (arrows) as the patient spreads the fingers open. In cases of chronic ulnar neuropathy, look for wasting of the first dorsal interossei (star).*

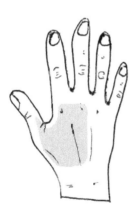

◀ **Radial Nerve** *sensory innervation (grey). A relatively small patch on the dorsum of the hand.*

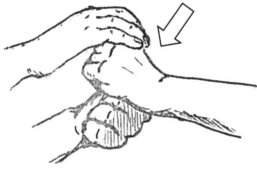

▶ **Radial Nerve** *motor exam, the radial n. innervates wrist and finger extension – test by supporting the hand and pushing resisting active extension (arrow).*

The Foot Sensory and Motor Exam

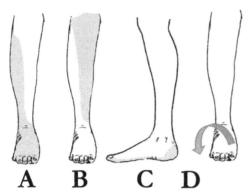

◀ *A) Common peroneal nerve sensory territory, B) femoral nerve sensory territory, C) tibial nerve sensory territory – predominantly the bottom of the foot, and D) foot eversion (shown) is weak in peroneal nerve palsy (i.e. foot drop) while foot inversion is spared – loss of both inversion and eversion suggests an L5 radiculopathy.*

Coordination

Coordination involves finely coordinated movements, including the ability to quickly and accurately hit a target or smoothly perform rapid movements. Coordination is most commonly tested by having a patient touch their nose, then reach out and touch the examiner's finger. The best exam is done by having the patient maximally extend their arm when they reach for the target. Abnormal coordination, termed *ataxia*, causes difficulty with smooth movements. On finger to nose testing patients will not follow a straight path when reaching for a target - their arm will swing side to side or up and down as they repeatedly veer off target and then overcorrect. They may reach too far, or not far enough, due to difficulty judging distance – a finding called *dysmetria*. You can also test coordination by asking patients to perform rapid movements such as finger tapping, or making and opening a fist quickly. Cerebellar dysfunction should *not* make a patient perform the finger to nose test in an excruciatingly slow fashion, as may be seen with weakness or psychogenic conditions.

Dysmetria and ataxia reflects usually underlying cerebellar dysfunction, or injury to cerebellar connections in the pons and basal ganglia. Profound proprioceptive loss seen in some thalamic or parietal lobe strokes can also cause dysmetria.

Reflexes

Deep tendon reflexes are automatic muscle responses to unexpected muscle stretch and help to maintain balance and posture. Reflexes are best obtained by directly striking the muscle tendon with a reflex hammer. The most commonly checked deep tendon reflexes include biceps, brachioradialis, triceps, patellar and achilles tendons, but in theory you can obtain a reflex from any muscle.

Recall that brisk reflexes may indicate an upper motor neuron injury, while decreased or absent reflexes may indicate lower motor neuron (i.e. a peripheral nerve) injury. Look for asymmetric reflexes, such as unilateral, or lower extremity rather than upper extremity hyperreflexia. In general, brisk reflexes are a finding that needs to be interpreted in the broader context of the patient's exam – however, some findings are clearly abnormal. Pathologic findings include sustained myoclonus, which is the non-stop beating of the foot when pressure is applied to the ankle, or an extensor (upgoing) great toe when the sole of the foot is lightly stroked. The presence of an upgoing toe is called the Babinski response. You can also say that the plantar reflex is flexor (normal), extensor (abnormal) or mute if there is no response.

Reflex Rating Scale

0 – No reflex
1+ – Requires activating maneuvers
2+ – Normal reflex
3+ – Spread to adjacent joint, brisk
4+ – Sustained clonus

If you have difficulty obtaining reflexes, you can perform an augmenting maneuver, where you distract the patient by asking them to perform tasks such as looking up, squeezing their fists or biting down, while you check their reflex. By distracting the patient you make it easier for automatic reflexes to manifest. Just be sure to perform a similar procedure on both sides in order to compare responses.

◀ *Extensor plantar reflex, with light stimulation to the bottom of the foot the great toe goes up – typically a sign of upper motor neuron dysfunction. It can be seen transiently after a seizure or general anesthesia.*

Gait

Testing gait is one of the most important parts of the general neurologic exam. It provides a rapid demonstration of balance, lower extremity strength, and gives clues about overall functional status. Although the various gait abnormalities will not be discussed here, the key pearl is to get into the habit of walking every patient when possible, and take note of posture, arm swing, gait width, balance and ability to turn smoothly. Station refers to the patients posture – normal, stooped, or hyperextended.

Notes

Paul D. Johnson, MD

Neuroimaging

"Surely you can't be serious."
- *"I am serious. And don't call me Shirley."*
Airplane!

Imaging is becoming increasingly important in the diagnosis and management of many neurologic illnesses. The neurologist and neurology APP must be competent in understanding when to order imaging tests, know the utility and limitations of such tests, be capable of interpreting radiology reports, and be able to independently evaluate common neuroimaging studies.

The four primary types of neuroimaging that APP's should be familiar with are computed tomography (CT), magnetic resonance imaging (MRI), catheter angiography and ultrasound. We will briefly review the relevant information for each of these modalities.

Computed Tomography

The head CT is easily the most frequently obtained neuroimaging study. It is widely available, rapidly acquired and relatively inexpensive. If no IV contrast is used, it is a non-invasive study. The CT machine essentially uses a series of X-rays to reconstruct an image of the brain. X-rays are sensitive to the density of materials, so bright structures on CT are 'hyperdense' and darker structures are referred to as 'hypodense.'

The head CT is very good at quickly identifying abnormal hyperdense lesions, such as blood (which is 'dense' because the hemoglobin in blood contains iron), foreign objects, calcified structures and bone fractures. It can also quickly identify very hypodense structures, such as free air in the brain, as sometimes happens following trauma or neurosurgery, or encephalomalacia.

Gross anatomy is moderately well visualized on head CT. Large abnormalities, such as prior areas of stroke that have had time to evolve into empty, cystic spaces are typically well visualized. Large tumors may be seen if they distort the shape of the brain, bleed or are calcified. However many tumors will not be obvious on head CT. Smaller, more subtle injuries in the brain, including most acute ischemic strokes, are also not well seen on CT. These cases may require brain MRI for better visualization.

A primary use of the head CT in the emergency room is to evaluate for hemorrhage – either within the brain tissue (intraparenchymal hemorrhage), between the brain and the dura matter (subdural hemorrhage), or between the brain and its immediate covering the pia matter (subarachnoid hemorrhage). An important point to remember is that in very early ischemic stroke, the head CT will usually be normal – it takes up to 6 hours for ischemic stroke changes to be reflected on a head CT. The main utility of the head CT in acute stroke is to rule out hemorrhage.

Drawbacks of head CT include the use of radiation, an especially important consideration in children or for those who may be pregnant. In addition, CT is prone to streak artifacts from very dense materials, such as the metal in dental fillings, EEG leads or the metal clips used to treat intracranial aneurysms. Although CT is rapidly acquired, motion artifact is also possible.

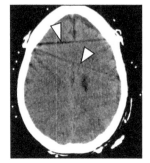

◄ *Non-enhanced head CT where the patient has EEG leads on, creating streak artifact (arrowheads). The same type of artifact is often caused by braces, dental work, or any other metallic object.*

Common uses of head CT include evaluating for hemorrhage or fracture, tracking post-operative changes, evaluating for brain swelling in ischemic stroke and as a rapid screening test in patients with sudden neurologic changes. Keep in mind that many inflammatory conditions, tumors and early strokes will be missed by a non-enhanced head CT.

A note on CT of the spine- although generally good at evaluating the vertebral bones for fractures or misalignment, the spinal cord itself is not well seen. For suspected spinal cord pathology, MRI is generally a better choice.

One other important aspect of CT is the ability to obtain reformats in different planes. Upon request, the radiology technologist can reform the head CT into coronal or sagittal views. This is done by the computer, and doesn't require any additional scanning or radiation for the patient, unlike brain MRI as we will see.

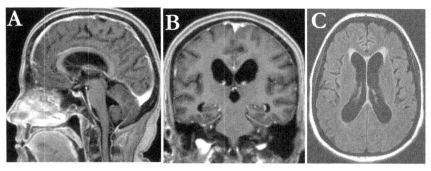

▲ *Brain MRI given in the A) sagittal view, B) coronal view, and C) axial view for the same patient.*

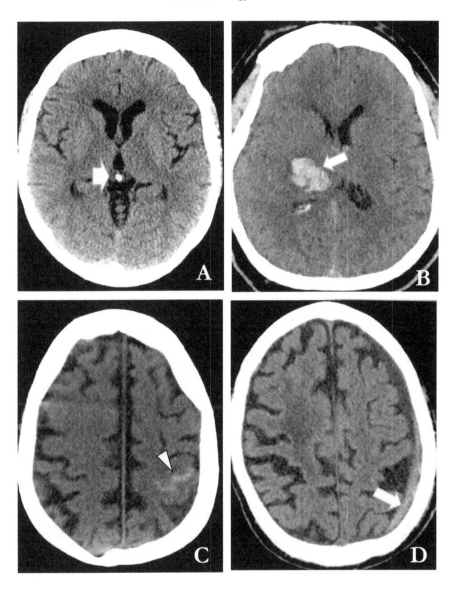

A) Normal non-contrast head CT. Note the calcified pineal gland in the center (short arrow). B) Intraparenchymal hemorrhage (arrow) in the right thalamus – this is a classic location for a hypertensive bleed. C) Small subarachnoid hemorrhage on the left (arrowhead). Subarachnoid and subdural hemorrhages are sometimes seen best in the coronal plane. D) A small left hemisphere subdural hematoma (arrow). See how it does not go into the brain sulci, whereas the subarachnoid hemorrhage did?

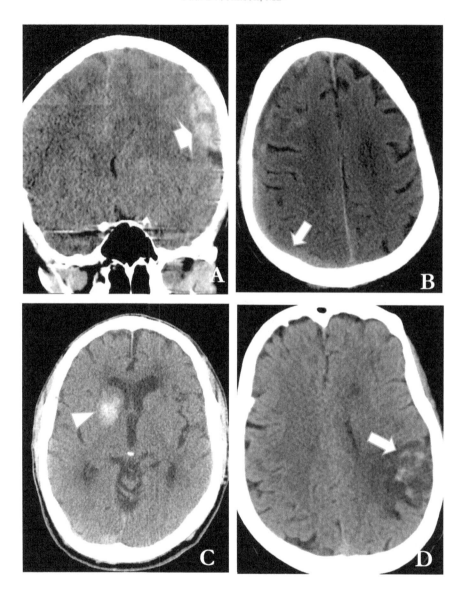

A) Left temporal epidural hematoma (short arrow), with some mass effect – note the bulging, convex appearance. B) Right subdural hematoma (arrow) – as the blood ages the density becomes increasingly similar to that of the brain tissue. C) Right caudate head intraparenchymal hemorrhage (arrowhead), likely from hypertension. D) Left temporal ischemic stroke with hemorrhagic conversion (arrow) – this represents bleeding into the dead tissue, and is much more likely to occur in embolic type strokes.

CT Angiography

An important variation of the head CT is the CT angiogram, which uses intravenous iodinated contrast to visualize the arteries in the head and neck. This can be extremely helpful in identifying vascular lesions such as carotid stenosis or dissections, acute arterial occlusion, aneurysm or vascular malformations. It is commonly used during acute stroke to evaluate for culprit vascular lesions or large vessel occlusions that may be amenable to intra-arterial thrombectomy. In addition, contrast may leak out of inflamed or fragile blood vessels in the brain, showing enhancement in inflammatory lesions such as brain abscesses or brain tumor.

The main limitation to using CT contrast is in individuals who have poor renal function. In general, CT contrast is not used in people who have a GFR less than 30 mL/min, although some exceptions do exist, such as dialysis dependent patients who can undergo dialysis after the study. Patients may also have allergic reactions to the contrast agent.

Radiologists will often report on any flow limiting areas of stenosis (narrowing) on the study, including the degree of narrowing of the proximal internal carotid artery (ICA). ICA narrowing is often measured using the North American Symptomatic Carotid Endarterectomy Trial (NASCET) criteria. The NASCET study indicated that in patients with minor stroke or TIA whose stroke was likely due to ipsilateral carotid artery disease, those with narrowing of 70% or more were likely to benefit from carotid endarterectomy.

Magnetic Resonance Imaging

MRI provides excellent visualization of the brain tissue, inflammation and stroke. Unlike CT, the MRI uses no radiation and does not measure tissue density. Instead, it depends on the magnetic property of the tissue – some people have called MRI a 'water map' of the brain because water creates such a strong signal. The MRI can be configured to highlight different properties of the brain, depending on the sequence used. Because you are not measuring density, we use the terminology "hyperintense" for bright structures and "hypointense" for darker structures on MRI. A major part of learning to interpret brain MRI is learning what brain features each distinct sequence is intended to highlight.

Unlike the head CT, brain MRI is very time intensive to acquire. Each individual sequence is obtained separately. So, the more information you need from the study, the longer it takes. Patients are required to stay very still, which can be challenging given the long duration of the study – often 30 to

60 minutes. It can also be difficult for claustrophobic patients, as the opening in the magnet is very small and the machine is loud.

In addition, the MRI is prone to artifact from any ferromagnetic material, which may include braces, metal screws, pins or clips and other material.

Let's review the most common brain MRI sequences. Bear in mind that individual hospitals and MRI machines may slightly differ in the names of some sequences. Use this section as a reference when needed.

T1-Weighted sequences typically give good anatomical representation of the brain. The cerebrospinal fluid (CSF) is dark, white matter appears white and grey matter appears grey. Inflammation and acute stroke do not show up well, but gadolinium based contrast will appear bright.

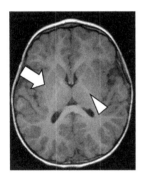

◀ *T1W brain MRI, providing a good view of the structural anatomy of the basal ganglia (arrow) and the internal capsule (arrowhead). Recall that the basal ganglia is made of the putamen and globus pallidus.*

T2-Weighted is a fluid bright sequence. This means, unlike T1W, the CSF appears bright. Areas of inflammation or injury will also be bright, although it can be hard to pick these out given the presence of bright CSF as well. In T2W images the white matter appears darker, and the grey matter appears brighter.

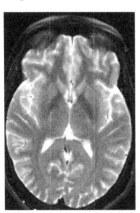

◀ *T2W brain MRI demonstrating 'bright' CSF. Notice that the white/grey matter are reversed compared to the T1W image above.*

T2W FLAIR is meant to make it easier to see pathology on a T2W sequence. Essentially, the bright signal from CSF is removed, leaving the pathologic signal from injured tissue clearly visible. This is often considered a "go-to" sequence for looking at pathology. Signs of chronic white matter disease (microvascular disease) is also readily apparent on T2W FLAIR.

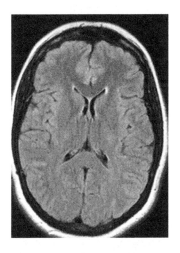

◄ *T2W FLAIR image, now with the bright CSF signal removed. This is an excellent sequence for looking at pathology – including subacute or chronic strokes, inflammation, white matter disease, etc. within the brain tissue.*

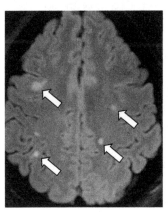

◄ *This T2W FLAIR image shows several hyperintense lesions (arrows), which are not specific, but could represent subacute strokes or areas of inflammation, such as in multiple sclerosis.*

Diffusion Weighted Imaging (DWI) is best for visualizing acute stroke. A stroke will appear as a bright lesion, which fades slowly over about two weeks. There are a few other things which will cause bright signal (called diffusion restriction) on DWI, but these are much less common – the pus in a brain abscess may cause diffusion restriction, as will some tumors. The Apparent Diffusion Coefficient (ADC) map is a sequence in which acute strokes appear dark – a bright lesion on both ADC indicates a chronic lesion.

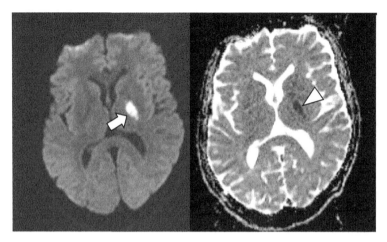

▲ *Diffusion Weighted Imaging (DWI) on the left, showing an internal capsule stroke (arrow). The ADC map on the right shows the stroke as a dark spot (arrowhead), verifying its acute nature.*

T2* this category includes several types of sequence, such as GRE, SWI or SWAN. All are similar, and show both blood and calcification. They are very sensitive, and are often used to identify *micro*-hemorrhages. These micro-hemorrhages may be too small to be seen even on CT scan. Both blood and calcification will appears as dark circles, which are larger than the underlying lesion.

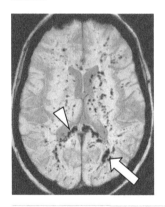

◀ *This T2* image (the category includes GRE and the more sensitive SWI sequence), shows diffuse microhemorrhage (arrow, but seen diffusely) within the white matter tracts. Microhemorrhage is common after high speed head trauma or extensive time on cardiac bypass. Note there is also hemorrhage within the corpus callosum (arrowhead).*

	Grey Matter	**White Matter**	**CSF**
T1W	Grey	White	Dark
T2W	White	Grey	Bright
FLAIR	White	Grey	Dark

▲ *Appearance of brain structures in the most common MRI sequences.*

MR Angiography

A quick note on angiography using MRI. As already mentioned, the contrast agent is gadolinium, not iodinated contrast (thus, safe for people with iodine allergies). Gadolinium is generally well tolerated, but has been known to cause a very serious, even lethal, condition known as nephrogenic systemic fibrosis in people with severe renal failure. Therefore, it should never be used in anyone with a GFR below 30 mL/min. It is also not safe to use gadolinium in pregnancy.

Gadolinium is often given when obtaining a brain MRI to evaluate for inflammatory conditions such as multiple sclerosis, or brain tumors. It is used for angiograms of the neck. However, for angiography of the head, gadolinium in not required. This is based on a neat trick based on MRI physics, because flowing blood is not well captured by the MRI scan, resulting in flow related absence of signal. These *flow voids* then are used to map out blood vessels. The results are best for angiography of the head; it can be done in the neck, but the results are poorer. This type of non-contrast angiogram is known as a "time of flight" study.

Conventional Catheter Angiogram

Another technique for imaging the vasculature of the head and neck is conventional catheter angiography, sometimes called digital subtraction angiography (DSA). This requires gaining arterial access with a micro-catheter, usually through the femoral artery. The catheter is then guided to the cerebral vasculature, at which point contrast dye is injected through the catheter and a series of images are taken. The resulting images are high quality, and show the flow of contrast (and thus the flow of blood) through the arteries, capillaries and veins in real time. This endovascular technique can be used to remove occlusive blood clots in acute stroke (thrombectomy), place a stent for carotid artery stenosis, or place wire coils to occlude an aneurysm.

Cather angiography provides the highest quality vascular imaging, but is an invasive technique and thus is done less frequently. It is not uncommon to have small, scattered micro-infarcts after catheter angiography due to the embolization of arterial plaque knocked loose by the catheter. However, symptomatic stroke as a result of catheter angiography is uncommon.

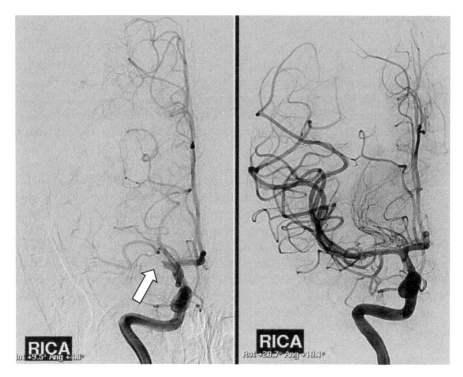

▲ *A conventional catheter angiogram of the right internal carotid artery, showing a proximal middle cerebral artery occlusion (left, arrow) followed by good perfusion after thrombectomy (right).*

Ultrasound

The most common ultrasound study used for neurology patients is carotid ultrasound, which is a non-invasive method of assessing for carotid stenosis. It can also be helpful in identifying carotid dissection, or tears in the wall of the artery. It is a relatively inexpensive test, but is performed less often now due to the frequency of obtaining CT angiograms.

Transcranial Doppler Ultrasound (TCD) is another common application of ultrasound. Specially trained TCD technicians can determine the direction and velocity of the flow of blood within the brain, carotid and vertebral arteries. It is especially useful following subarachnoid hemorrhage, where spasm of the cerebral arteries may cause significant increases in the velocity of blood flow in affected area. TCD can also be used to evaluate for active embolization, say from a symptomatic carotid artery plaque. However, the availability of TCD is limited and not readily available in all hospitals.

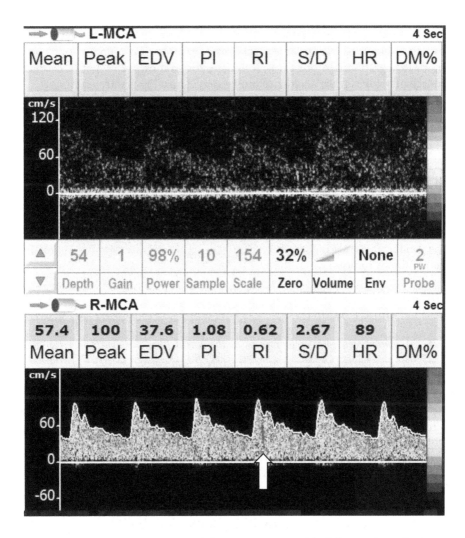

▲ *Transcranial Doppler Ultrasound showing waveform of blood flow in the cerebral circulation. An emboli (arrow) was detected. The most common use is in detecting vasospasm after subarachnoid hemorrhage.*

Echocardiography

Although not a neuro-diagnostic study per-se, echocardiograms are frequently ordered in people suspected of having ischemic strokes. The echocardiogram is used to evaluate for a number of stroke risk factors, such as presence of thrombus which could be the source of the stroke, for atrial dilation indicating a high risk for atrial fibrillation, or patent foramen ovale (PFO) which could be a source of thromboembolism. The majority of stroke patients undergo transthoracic echocardiography only. Transesophageal echocardiography is more invasive, but is sometimes used when a cardiac source of stroke is suspected and transthoracic imaging is unrevealing.

Venography

Sometimes we want to take a close look at the veins, not just the arteries, around the brain. This is primarily to evaluate for dural sinus thrombosis – essentially a DVT surrounding the brain. It is also used for certain brain tumors, when they are in close proximity to the cerebral veins. Venography can be done with CT or MRI.

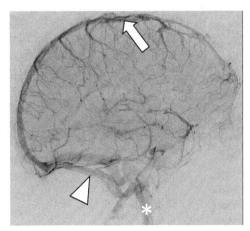

▲ *A conventional catheter angiogram in the venous phase, demonstrating the superior sagittal sinus (arrow), transverse sinus (arrowhead), and internal jugular veins (*).*

All the rest

There are many other imaging studies in the field of neurology, including nuclear medicine perfusion studies, DAT scans, CSF flow studies, other MRI sequences, CT myelograms and others. However, they are outside the scope of this book and uncommonly used.

"It be iddy biddy baby doo doo"

The mnemonic 'It be iddy biddy baby doo doo' can be helpful when figuring out the age of blood on a brain MRI, as the pattern of intensity changes quickly as the blood products break down. The following table is a reference:

Mnemonic	Stage	T1W	T2W
It Be	Hyperacute	Isointense (I)	Hyperintense (B)
IdDy	Acute	Isointense (I)	Hypointense (D)
BiDdy	Early Subacute	Hyperintense (B)	Hypointense (D)
BaBy	Late Subacute	Hyperintense (B)	Hyperintense (B)
Doo Doo	Chronic	Hypointense (D)	Hypointense (D)

(I) Isointense; (B) bright or hyperintense; (D) dark or hypointense. Table refers to age of BLOOD only.

Notes

Neuro-Imaging Cases

"I am Groot."
Guardians of the Galaxy

Case 1 – Neuro-Imaging

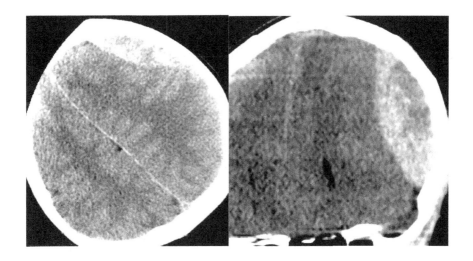

A 15 year old boy fell off his bike with no helmet and struck his head. He felt fine for twenty minutes before becoming lethargic and difficult to arouse.

1. What is the diagnosis?
A. Subdural hematoma
B. Epidural hematoma
C. Meningioma
D. Abscess

2. This condition is typically caused by an injury to which blood vessel(s)?
A. Temporal artery
B. Middle cerebral artery
C. Middle meningeal artery
D. Bridging veins

3. What is the primary treatment of this condition?
A. Steroids
B. Mannitol or hypertonic sodium chloride
C. Surgical drainage
D. Blood pressure control

Case 1 – Neuro-Imaging

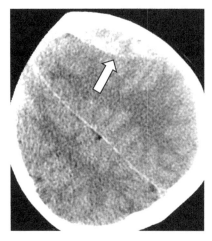 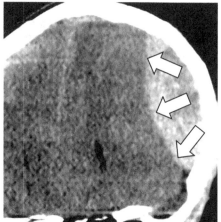

This is a classic epidural hematoma (arrows), usually associated with a skull fracture and damage to an artery – specifically, the middle meningeal artery. Blood accumulates between the skull and the dura, expands quickly, and causes brain herniation and possibly death if not treated emergently with surgical drainage.

Subdural hemorrhages can also cause significant mass effect, but are concave, not convex like an epidural bleed. Because subdural hemorrhages are caused by tears to lower pressure bridging veins they do not typically expand quickly, unless the patient has a coagulation disorder.

Mannitol or hypertonic saline are used for stroke associated edema leading to herniation.

1. What is the diagnosis?
B. Epidural hematoma – Note the convex shape. These bleeds are contained by dural suture lines, unlike subdural bleeds.

2. Injury to which blood vessel is most commonly responsible?
C. Middle meningeal artery – a small extracranial artery that enters the skull at the temple

3. What is the primary treatment of this condition?
C. Surgical drainage – this is a surgical emergency

Case 2 – Neuro-Imaging

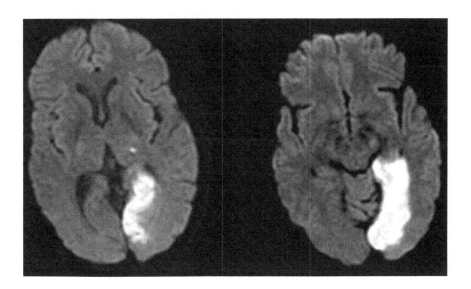

Answer the questions based on this patient's brain MRI.

1. What MRI sequence is shown here?
A. T2-Weighted FLAIR
B. Diffusion Weighted
C. T1-Weighted post-contrast
D. T2-Weighted

2. In which vascular territory is the pathology?
A. Anterior cerebral artery
B. Middle cerebral artery
C. Posterior cerebral artery
D. Middle meningeal artery

3. Which best describes the patient's most likely clinical symptoms?
A. Loss of vision
B. Loss of speech
C. Hemiparesis
D. Loss of sensation

Case 2 – Neuro-Imaging

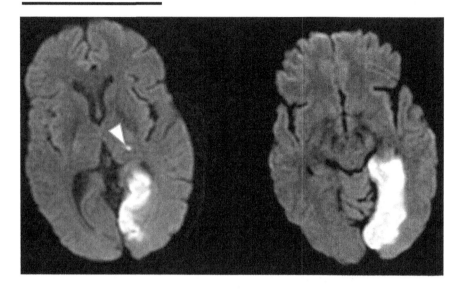

This patient has a large stroke of the left occipital and medial temporal lobe (as well as a very small stroke in the right thalamus, shown by the arrowhead). This is in the posterior cerebral artery territory – recall that branches of this artery also supplies the thalamus. The occlusion of a large artery is suspicious for embolism. Stroke of the left lower occipital lobe will likely cause a right upper visual field loss in both eyes.

1. What MRI sequence is shown here?
B. Diffusion Weighted

2. In which vascular territory is the pathology?
C. Posterior cerebral artery

3. Which best describes the patient's most likely clinical symptoms?
A. Loss of vision – it partly affects the occipital lobe

▶ *This patient's formal visual field testing, demonstrating a right upper visual field cut, as expected.*

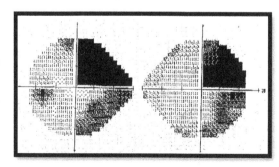

Case 3 – Neuro-Imaging

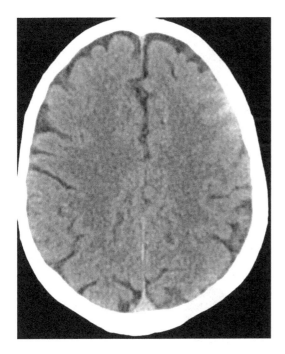

Answer the questions based on this patient's imaging shown above

1. Which description best fits this CT scan?
A. Normal non-enhanced head CT
B. Abnormal non-enhanced head CT
C. Normal post-contrast head CT
D. Abnormal enhanced head CT

2. What abnormality, if any, is present?
A. Normal head CT
B. Subacute infarct present
C. Subdural hematoma present
D. Subarachnoid hemorrhage present

Case 3 – Subarachnoid hemorrhage

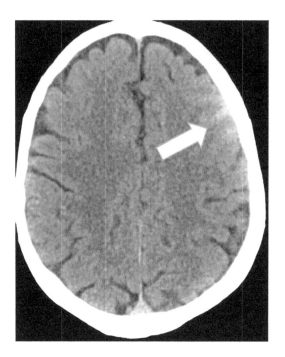

This is a non-enhanced head CT – there is no contrast in the blood vessels or sinuses. However, there is an abnormal hyperdense signal within the sulci of the left frontal lobe (arrow). Compare this to the opposite side, where you see dark CSF density signal within the sulci. This image is diagnostic of a small subarachnoid hemorrhage – they can be subtle. Compare to the subarachnoid in image C on page 42.

1. Which description best fits this CT scan?
B. **Abnormal non-enhanced head CT due to hemorrhage**

2. What abnormality, if any, is present?
D. **Subarachnoid hemorrhage present. This might be seen well in a coronal reformat.**

Case 4 – Neuro-Imaging

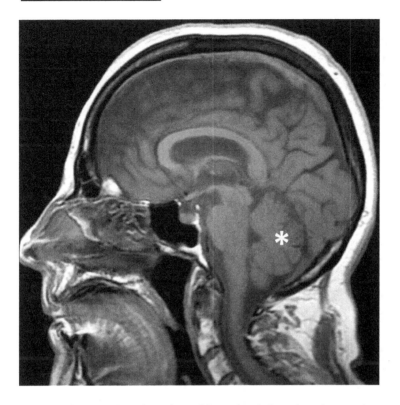

Answer the questions based on this patient's imaging shown above.

1. Which MRI sequence is shown?
A. T1-Weighted
B. T1-Weighted post-contrast
C. T2-Weighted
D. T2-Weighted FLAIR

2. Which brain structure is represented by the * ?
A. Cerebellum
B. Fourth ventricle
C. Corpus callosum
D. Occipital lobe

3. In this imaging sequence, the cortex is what color?
A. White   B. Black
C. Grey   D. Not visible

Case 4 – Neuro-Imaging

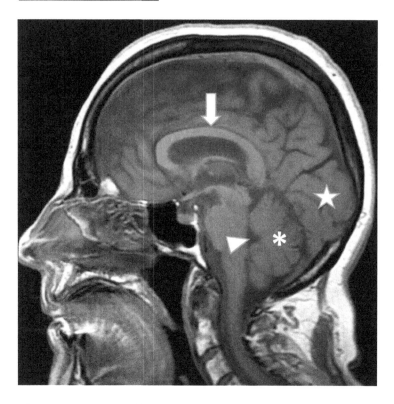

This is a T1-Weighted sagittal brain MRI. The corpus callosum (arrow) is a white matter tract – on T1W images the white matter appears white and the grey matter (i.e. the cortex) is grey. On T2W imaging this is inverted. The fourth ventricle is indicated with an arrowhead, the occipital lobe with a star.

1. Which MRI sequence is shown?
A. T1-Weighted imaging

2. Which brain structure is represented by the * ?
A. Cerebellum

3. In this imaging sequence, the cortex is what color?
C. Grey – and white matter appears white

> **Teaching Point:** T1W images do a good job of showing brain anatomy, but T2W FLAIR is best for seeing pathology!

Case 5 – Neuro-Imaging

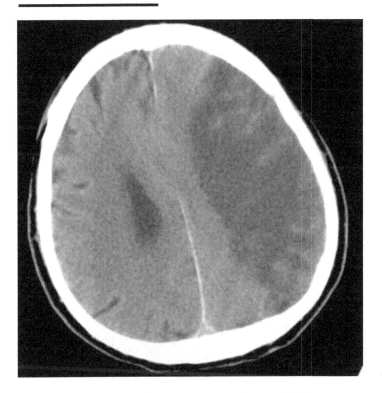

A 58 year old man is admitted to the stroke ward. His non-enhanced brain CT is shown above.

1. What is the most likely timing of this CT scan, in relation to the patient's stroke?
A. Within 60 minutes of stroke onset
B. Within 6 hours of stroke onset
C. Within 2-3 days of stroke onset
D. Approximately 4 weeks after stroke onset

2. Which brain vascular territory is affected?
A. Anterior cerebral artery
B. Middle cerebral artery
C. Posterior cerebral artery
D. Basilar artery

Case 5 – Neuro-Imaging

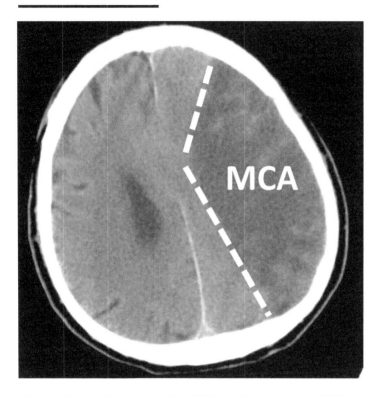

This patient had a large left middle cerebral artery (MCA) territory stroke, due to untreated atrial fibrillation. A non-enhanced CT scan generally takes at least 6 hours to show evidence of acute ischemia. This much frank infarct, with evidence of swelling and mass effect is more likely to be seen within 2-3 days after stroke onset. Notice the lack of sulci within the stroke bed, as compared to the normal side, which is due to swelling. The MCA vascular territory is clearly defined.

1. What is the most likely timing of this CT scan, in relation to the patient's stroke?
C. Within 2-3 days of stroke onset, giving time for peak edema to develop

2. Which brain vascular territory is affected?
B. Middle cerebral artery – this image nicely demonstrates a stroke of the entire MCA territory

Case 6 – Neuro-Imaging

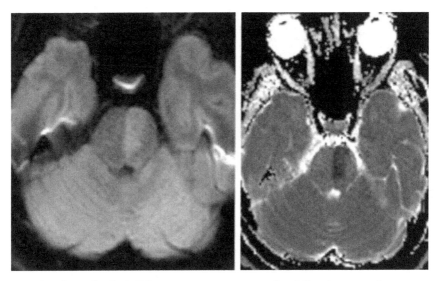

Use the above brain MRI sequences to answer the following questions.

1. Based on the MRI, how long ago did the patient's stroke occur?
A. Within the past 10 days
B. Between 2-4 weeks ago
C. More than 4 weeks ago
D. Timing of the stroke cannot be determined from this MRI

2. Which brain region is affected by the stroke?
A. Midbrain
B. Pons
C. Medulla
D. Cerebellum

3. What symptoms do you expect this patient to have clinically?
A. Inability to understand language
B. Right hemiparesis
C. Right visual field loss
D. Loss of sensation in both legs

Case 6 – Neuro-Imaging

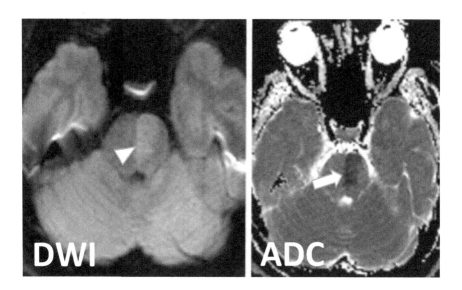

This patient has a left pontine stroke (arrowhead). The cause is likely atherosclerotic disease of a penetrating branch of the basilar artery. Notice how the stroke respects midline – separate basilar artery branches supply each side of the pons. We know the stroke is relatively recent, no more than 10-14 days old, because the ADC map remains dark (arrow). After 2 weeks this darkness will normalize and disappear. A pontine stroke will cause contralateral weakness. You would also expect to see cranial nerve dysfunction, possibly swallowing problems or eye movement abnormalities.

1. Based on the MRI, how long ago did the patients stroke occur?
A. Within the past 10 days – the ADC normalizes after that time

2. Which brain region is affected by the stroke?
B. Pons – this is the largest part of the brainstem

3. What symptoms do you expect this patient to have clinically?
B. Right hemiparesis – remember that the left brain motor fibers cross to innervate the right side of the body. However, this patient may have left facial weakness as well, if the left facial nerve (CN VII) is affected as it leaves the pons.

Case 7 – Neuro-Imaging

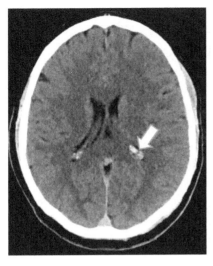
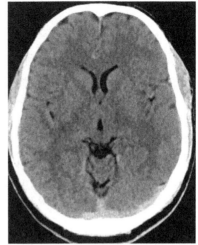

The images above were obtained from a 35 year old man who had chronic headaches.

1. What type of images are these?
A. Non-enhanced head CT
B. CT with iodine based contrast
C. CT with gadolinium contrast
D. CT angiogram

2. Which of the following best describes the white structure indicated with an arrow?
A. Benign calcification
B. Tumor
C. Hemorrhage
D. Cannot be identified from these images

3. What is the major finding of these images?
A. Normal study
B. Hydrocephalus
C. Ischemic stroke
D. Hemorrhagic stroke

Case 7 – Neuro-Imaging

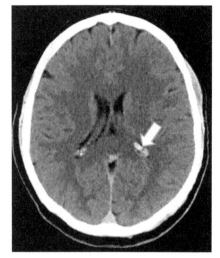
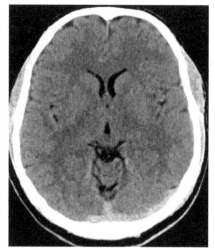

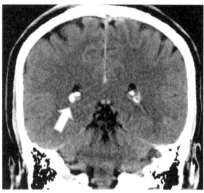

These are both normal non-enhanced head CT images from a young adult. There is calcification of the choroid plexus, which is common – a coronal view is shown here as well, to better visualize the calcified choroid (arrows). Recall that the choroid plexus is responsible for producing cerebrospinal fluid. Common areas for benign calcification include the dura, the basal ganglia and the pineal gland.

1. What type of images are these?
A. Non-enhanced head CT – there is no contrast in these images. Contrast is used to obtain a CT angiogram or CT venogram, in which case we would see enhancement within the vasculature

2. Which of the following best describes the white structure indicated by an arrow?
A. Benign calcification – this is benign calcification of the choroid plexus, which is common and the incidence increases with age

3. What is the major finding of these images?
A. Normal study – these are normal head CT images from a young adult. Note that the brain looks more 'full' than it might for someone several decades older

Notes

Paul D. Johnson, MD

Stroke

"Do you know the 'f' word?"
- "Ffff...fornication?"
The King's Speech

Acute stroke is a medical emergency and requires rapid assessment, diagnosis and treatment. Evaluating for stroke is a significant component of the workload in inpatient neurology. Stroke is a broad term, referring to several types of cerebrovascular disease (i.e. dealing with the blood vessels of the brain). Typically, disease in specific blood vessels causes injury to the portion of the brain they supply – therefore disease in particular vessels typically causing focal brain injury and focal neurologic symptoms.

The management of stroke can vary greatly depending on the type of stroke, stroke etiology and the elapsed time from stroke onset. We will cover the basics for the most common stroke types here.

Ischemic stroke is the most frequent stroke type, accounting for about 85% of all strokes. Almost 800,000 Americans have strokes each year, making this a very common neurologic emergency. Ischemia occurs when blood flow is decreased or stopped altogether, resulting in the death of brain tissue. Ischemia typically occurs because of chronic atherosclerosis causing stenosis and thrombosis within a blood vessel, or from blood clots that form elsewhere in the body and travel (embolize) to the brain vasculature and block blood flow in the vessel they occlude. A major cause of embolic stroke is cardiac embolism from atrial fibrillation.

Hemorrhagic strokes are less common, but account for a large number of stroke related deaths. These are bleeding type strokes, typically due to the rupture of a small blood vessel, made fragile over time due to long standing high blood pressure. The result is a hematoma within the brain tissue, known as an intraparenchymal hemorrhage (IPH).

Another type of hemorrhagic stroke is subarachnoid hemorrhage, in which blood fills the spaces around the brain. These usually occur from ruptured aneurysms, fragile outpouchings which grow from a weak spot in a blood vessel. Aneurysm usually develop at the junction of two blood vessels. These are typically managed by neurosurgeons, and will be discussed in more detail in the neurosurgical section of the book.

Transient Ischemic Attack (TIA) is a stroke like episode with spontaneous resolution. In the past a TIA was defined as a stroke like episode that resolved completely within 24 hours. However, now that MRI is easily accessible, a TIA is defined as a stroke like event that leaves no evidence of stroke on MRI. Typically these will last 5 – 15 minutes, but can be shorter or longer. A TIA is a warning sign for impending stroke, and should be taken very seriously. A full stroke evaluation is indicated, including evaluation for starting medication for primary stroke prevention, such as aspirin or statins.

The most commonly used exam for the evaluation of a patient with a known or suspected stroke is the National Institutes of Health Stroke Scale, or **NIHSS**. The NIHSS is a standardized, widely used exam whose purpose is to help determine the severity of stroke symptoms. It was not designed to make the diagnosis of stroke, and a high stroke scale does not mean stroke is more likely – rather, the more severe the stroke, the higher the patient scores on the NIHSS.

The American Heart Association offers an online NIHSS certification: https://learn.heart.org/Activity/2695217/Detail.aspx

Statins and Ischemic Stroke
How are statins used in stroke? First, they are used in ischemic strokes, not hemorrhagic stroke – as very low lipids are thought to increase the risk of bleeding. Secondly, there is evidence that high dose statins – usually atorvastatin 40 mg – 80 mg, reduces the risk of recurrent stroke and other vascular disease. It is standard practice to start a high dose statin, like atorvastatin or rosuvastatin after ischemic stroke or TIA, unless a contraindication exists.

Category	Score		Time	Score
1a. Level of Consciousness (LOC) (Alert, drowsy, etc.)	0 = 1 = 2 = 3 =	Alert Drowsy Stuporous Coma		
1b. LOC Questions (Month, age)	0 = 1 = 2 =	Answers both correctly Answers one correctly Incorrect		
1c. LOC Commands (Open/close eyes, make fist & let go)	0 = 1 = 2 =	Obeys both correctly Obeys one correctly Incorrect		
2. Best Gaze (Eyes open - pt follows examiner's fingers or face)	0 = 1 = 2 =	Normal Partial gaze palsy Forced deviation		
3. Visual (Introduce visual stimulus/threat to pt's visual field quandrants. Cover 1 eye and hold up fingers in all 4 quadrants.)	0 = 1 = 2 = 3 =	No visual loss Partial hemianopsia Complete hemianopsia Bilateral hemianopsia		
4. Facial Palsy (Show teeth, raise eyebrows and squeeze eyes tightly shut.)	0 = 1 = 2 = 3 =	Normal Minor Partial Complete		
5a. Motor Arm - Left (Elevate extremity to 90 degrees and score drift/ movement. Count to 10 out loud and use fingers for visual cue.)	0 = 1 = 2 = 3 = 4 = NT=	No drift Drift Can't resist gravity No effort against gravity No movement Amputation, joint fusion (Explain)		
5b. Motor Arm - Right (Elevate extremity to 90 degrees and score drift/ movement. Count to 10 out loud and use fingers for visual cue.)	0 = 1 = 2 = 3 = 4 = NT=	No drift Drift Can't resist gravity No effort against gravity No movement Amputation, joint fusion (Explain)		
6a. Motor Leg - Left (Elevate extremity to 30 degrees and score drift/ movement. Count to 5 out loud and use fingers for visual cue.)	0 = 1 = 2 = 3 = 4 = NT=	No drift Drift Can't resist gravity No effort against gravity No movement Amputation, joint fusion		
6b. Motor Leg - Right (Elevate extremity to 30 degrees and score drift/ movement. Count to 5 out loud and use fingers for visual cue.)	0 = 1 = 2 = 3 = 4 = NT=	No drift Drift Can't resist gravity No effort against gravity No movement Amputation, joint fusion (Explain)		
7. Limb ataxia (Finger to nose, heal down shin)	0 = 1 = 2 =	Absent Present in one limb Present in two limbs		
8. Sensory (Pin prick to face, arms, trunk, and legs -compare sharpness side to side, or no feeling at all.)	0 = 1 = 2 =	Normal Partial loss Severe loss		
9. Best Language (Name items, describe picture, and read sentences. Don't forget glasses if they normally wear them.)	0 = 1 = 2 = 3 =	No aphasia Mild to moderate aphasia Severe aphasia Mute		
10. Dysarthria (Evaluate speech clarity by pt reading or repeating words on list.)	0 = 1 = 2 = NT	Normal articulation Mild to moderate dysarthria Near to unintelligible or worse Intubated or other physical barrier		
11. Extinction and Inattention (Use information from prior testing or double simultaneous stimuli testing to identify neglect. Face, arms, legs and visual fields.)	0 = 1 = 2 =	No neglect Partial neglect Complete neglect		
NT= Not Testable acceptable as noted above				
TOTAL SCORE:				

▲ *The National Institutes of Health Stroke Scale (NIHSS) form.*

Another commonly used tool to assess the risk of recurrent stroke after a TIA is the **ABCD2 score**. This simple calculation helps predict a patient's chances for a recurrent stroke in the next 2, 7 and 90 days, based on clinical features at the time of presentation. The score is just a guide to help triage patients with TIA, and should not replace clinical judgement! As a reminder, TIA's are defined as focal neurologic symptoms due to cerebrovascular disease that resolve completely, without evidence of a stroke on brain MRI.

Many convenient ABCD2 calculators can be found online or through phone based applications. However, the components of the score include:

The ABCD2 score: higher scores mean higher risk of recurrent stroke.

Age > 60 years	+1 point if present
Blood Pressure > 140/90 mmHg (either SBP > 140 or DBP > 90)	+ 1 point if present
Clinical features of the TIA: - Unilateral Weakness - Speech disturbance, no weakness - Other symptoms	+2 points +1 point No points
Duration of symptoms - < 10 minutes - 10-59 minutes - > 60 minutes	No points +1 point +2 points
History of diabetes	+1 point

You know how.

Down to earth.

I got home from work.

Near the table in the dining room.

They heard him speak on the radio last night.

MAMA

TIP – TOP

FIFTY – FIFTY

THANKS

HUCKLEBERRY

BASEBALL PLAYER

◀ *Part of the NIH stroke scale, for testing language (aphasia) and articulation (dysarthria).*

Diagnostic Principles for Ischemic Stroke and TIA

The following are general rules of thumb and should be tailored to the individual patient.

Brain Imaging – patients usually undergo both non-contrast CT and MRI.
- **Head CT** – The initial study for acute stroke because it is fast and readily available. Its main uses are identifying hemorrhagic strokes, as there is no clinical way to differentiate between hemorrhagic and ischemic strokes. Large strokes may be visible, but small or acute strokes can be missed.
- **Brain MRI** – Typically done within 24 hours of presentation for acute stroke. MRI is best for identifying small strokes, and helps differentiate between TIA's and strokes whose symptoms resolve quickly.

Vessel Imaging – It is generally not necessary to obtain both a CTA and MA. Head and neck imaging is needed for all ischemic stroke patients, whereas head imaging only is needed for hemorrhagic strokes.
- **CT angiogram** – Often performed at initial presentation (i.e. in the ER) to evaluate for acute large vessel occlusion, symptomatic carotid stenosis, or large artery atherosclerosis.
- **MR angiogram** – A non-contrast MRA of the head, known as a "time of flight" study, is available for those who need to avoid iodinated contrast.
- **Carotid duplex** – Largely replaced by CTA or MRA, unless explicitly needed to evaluate high grade carotid stenosis prior to surgery.

Cardiac Monitoring
- **Telemetry** – Guidelines stipulate that patients should undergo at least 48 hours of cardiac telemetry to evaluate for occult atrial fibrillation as a cause of stroke. Most undergo much longer monitoring, especially if embolism is suspected.
- **Outpatient cardiac telemetry** – If occult atrial fibrillation is a possible stroke mechanism and isn't detected as an inpatient, 14-30 day outpatient cardiac monitors are frequently used. Implantable devices for even longer monitoring are now available.
- **Echocardiogram** – A transthoracic echocardiogram is a standard part of the stroke/TIA evaluation, usually performed with agitated saline (called a "bubble study") to evaluate for PFO.

Laboratory Evaluation

- **Lipid panel** – Does not need to be fasting. LDL is the most relevant result for stroke and TIA, and generally should be < 90 mg/dL or, for diabetics < 70 mg/dL.
- **HbA1c** – Should be checked in all stroke and TIA patients as poorly controlled diabetes is a modifiable stroke risk factor.
- **Troponin** – Should be checked in all acute stroke patients.
- **BNP** – An elevated BNP (>200) may be a sign of occult atrial fibrillation. Typically not needed in patients with known congestive heart failure or atrial fibrillation.
- **Hypercoagulation labs** – Hypercoagulable work-up is typically reserved for people of young age (i.e. < 55 years) with unexplained stroke, especially those with cerebral venous sinus thrombosis.

Hypercoagulable Lab Evaluation

Venous
- ✓ Activated Protein C Resistance → reflex to Factor V Leiden
- ✓ Prothrombin G20210A gene mutation
- ✓ Protein C and Protein S deficiency
- ✓ Antithrombin III deficiency
- ✓ Methyltetrahydrofolate Reductase (MTHFR)
- ✓ Antiphospholipid Antibodies

Arterial

Remember: All ischemic stroke patients should be discharged on an antiplatelet agent and a statin – if not, then document why not. The same applies to evaluation by physical, occupational and speech therapists. Any stroke patient with a facial droop or dysarthria should be NPO until cleared by a formal swallow evaluation due to the elevated risk for aspiration.

Case 1 – Stroke

A 64 year old, right handed woman is brought to the emergency room by her daughter. The patient was in the kitchen when she became weak in the left hand, dropping a plate. She wasn't able to move the left side of her mouth, her left foot dragged and her voice sounded slurred. The symptoms had resolved completely by the time medics arrived 5 minutes later, although her blood pressure was 172/104 mmHg. The patient came to the ER where her blood pressure had returned to normal, and her NIHSS showed no deficits, including normal left sided strength. The patient's history was notable only for high blood pressure, for which she took Lisinopril, and type 2 diabetes.

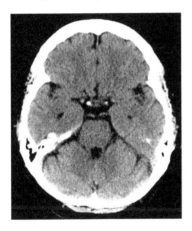

◄ *The non-enhanced head CT was read by the radiologist as showing "No acute intracranial abnormality," meaning there is no hemorrhage and no large, obvious stroke.*

1. Is the normal head CT enough to rule out an acute ischemic stroke?
A. Yes   C. Only when done with IV contrast
B. No

2. What is this patient's ABCD2 score?
A. 1   D. 4
B. 2   E. 5
C. 3   F. 6

3. The patient only takes Lisinopril. Which of the following medications should be started to reduce the risk of recurrent stroke?
A. Metoprolol 20mg BID
B. Warfarin 5mg
C. Aspirin 81mg
D. Apixaban 5mg BID

Case 1 – Diagnosis: TIA

This patient most likely had a transient ischemic attack, as the symptoms resolved within 5 minutes. However, a brain MRI would be necessary to fully exclude a small stroke – the head CT rules out hemorrhage, but does a poor job of evaluating acute ischemic stroke.

A transient ischemic attack should be taken as a high risk sign for impending stroke – the ABCD2 score of 5 suggests a 4% risk of stroke within 48 hours and almost 10% risk of stroke within three months. Clearly, preventative medications such as aspirin and statins are needed.

All ischemic stroke and TIA patients should be on an antiplatlet agent – either aspirin, clopidogrel or combination aspirin with extended-release dipyridamole (aggrenox) – unless contraindications exist. Antiplatelets reduce recurrent stroke risk by about 25%. Anticoagulants are used to prevent strokes in patients with atrial fibrillation. If someone has a TIA while already on an antiplatelet agent, the evidence is less clear about how best to treat them. Many providers will change from one antiplatelet agent to another. Research is ongoing about the benefit of short term use of two antiplatelet agents.

Terminology
Hemiparesis – Mild to moderate weakness of one half of the body
Hemiplegia – Complete paralysis of one half of the body

Teaching Point: Pure motor strokes causing isolated hemiparesis are frequently caused by small vessel disease involving the basal ganglia or thalamus, less frequently in the descending motor fibers in the pons.

1. Is the normal head CT enough to rule out an acute ischemic stroke?
B. No – acute stroke is usually not seen on head CT, you need MRI

2. What is the patient's ABCD2 score?
E. 5 (age, BP > 140/90, unilateral weakness, and history of diabetes)

3. The patient only takes Lisinopril. Which other medication should be started to reduce the risk of recurrent stroke?
C. Aspirin 81 mg - alternatively clopidogrel 75 mg could be used

Case 2 – Stroke

A 67 year old right handed man has a history of high blood pressure and hyperlipidemia. He has been on aspirin 81mg, rosuvastatin 10mg and hydrochlorothiazide 25mg only. Early in the morning he developed sudden onset of isolated right arm weakness. He arrives in the ER two hours after symptom onset, where his NIHSS was 3, for isolated right arm weakness. His blood pressure was 212/120 mmHg. He had no prior history of stroke, TIA or heart disease.

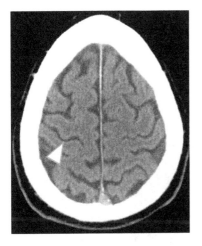

▲ The non-enhanced head CT showed no hemorrhage, no prior strokes and no overt abnormalities. The cortical motor strip, and specifically the so called 'hand knob' (arrowhead) where hand motor function is located, is well visualized.

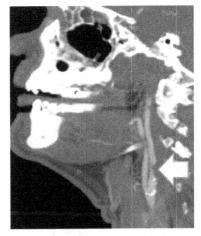

▲ The CT angiogram shows a large filling defect (arrow) in the proximal internal carotid artery, with just over 70% narrowing of the lumen of the artery.

1. What is the next best step in this patient's treatment?
A. Administer IV tPA
B. Administer aspirin 325mg
C. Administer IV labetalol
D. Emergency carotid endarterectomy

2. Which of the following is a likely location for this stroke?
A. Left occipital lobe
B. Right cortical motor strip
C. Right parietal lobe
D. Left cortical motor strip

Case 2 – Stroke

This patient received IV tPA after his blood pressure was controlled with IV labetalol. Uncontrolled hypertension is a strict contraindication to giving IV tPA because of the risk of brain hemorrhage. Blood pressure must be lowered to < 185/110 mmHg prior to giving tPA and kept below that level for the next 24 hours. Given the high grade (>70%) carotid artery stenosis on the symptomatic side, the stroke mechanism is presumed thromboembolism from ruptured carotid plaque.

1. What is the next best step in this patient's treatment?
C. Administer IV labetalol – blood pressure must be brought down to 185/110 mmHg or less before giving IV tPA

2. Which of the following is a likely location for this stroke?
D. Left cortical motor strip – carotid stenosis leads to embolism which most often affects the cortex

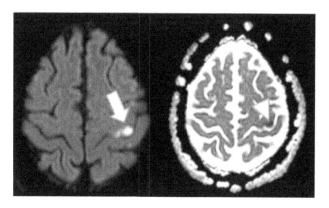

▲ *Brain MRI showed diffusion restriction (arrow) in the left motor strip, in the hand area, which was confirmed on the ADC (arrowhead). Note that cortical motor strokes can cause pure motor weakness!*

3. This patient had a left MCA stroke with high grade (>70%) left internal carotid artery stenosis, and was scheduled for left carotid endarterectomy. Which of the following patients would be <u>least</u> likely to benefit from carotid endarterectomy?
A. The patient from this case, undergoing surgery 1 week after stroke
B. The patient from this case, undergoing surgery 3 months after stroke
C. An asymptomatic patient with 75% carotid artery stenosis
D. A symptomatic patient with 30% carotid artery stenosis

Case 2 – Diagnosis: Symptomatic Carotid Artery Stenosis

3. This patient had a left MCA stroke with high grade (>70%) left internal carotid artery stenosis, and was scheduled for left carotid endarterectomy. Which of the following patients would be least likely to benefit from endarterectomy?
D. A symptomatic patient with 30% carotid artery stenosis

Carotid endarterectomy (CEA) and carotid stenting are frequently performed for patients with symptomatic internal carotid artery stenosis, causing TIA or minor stroke. Asymptomatic patients with high grade stenosis may be considered for CEA if their life expectancy is long and the surgeon performing the procedure has a record of very low mortality rates.

Teaching Point: CEA should be done between 2 days and 2 weeks of symptoms for the biggest risk reduction – but can be done up to 6 months out. The benefit is less if stenosis is 50-69%, and women in this group don't benefit at all. There is no benefit to CEA for carotid stenosis < 50%.

For more information, the AAN has published guidelines for the use of CEA in symptomatic and asymptomatic patients:
http://tools.aan.com/professionals/practice/guideline/pdf/Clinician_guideline.pdf

Complications of Carotid Endarterectomy
- **Hyperperfusion:** after surgically opening the artery and restoring blood flow there is a risk of the brain receiving more blood than it can handle – this can cause ipsilateral headache, seizure and hemorrhagic stroke. Patients must be watched closely after CEA for hypertension and development of these symptoms.
- **Acute stroke:** there is a small chance of blood clots forming in the carotid artery after the surgery and causing new stroke.
- **Nerve injury:** rarely the vagus, facial or sympathetic nerves which run within the carotid sheath are injured.

Case 2 – Diagnosis: Symptomatic Carotid Artery Stenosis

This patient had an excellent outcome after receiving IV tPA, with significant, but not complete, improvement in his arm strength. He underwent left carotid endarterectomy 3 days later, without complication.

IV tPA Inclusion & Exclusion Criteria – Partial List

Inclusionary Criteria

- Diagnosis of ischemic stroke, measurable neurological deficit
- Onset of symptoms within 4.5 hours
- Age at least 18 years

Exclusionary Criteria

- Evidence of hemorrhage on CT scan
- SBP > 185 or DBP > 110 mmHg and unresponsive to medication
- Platelet count below 100,000
- Use of warfarin with INR > 1.7
- Use of heparin with prolonged PTT
- Glucose < 50 mg/dL
- Severe head trauma, intracranial/spine surgery within 90 days
- Aortic arch dissection
- Vascular malformations, unruptured intracranial aneurysms > 10mm
- Recent GI or urinary tract hemorrhage within 21 days

Cautionary Criteria

- Ischemic stroke within prior 90 days
- Major surgery or trauma within 14 days
- Recent MI, depending on location
- Within the 3-4.5 hour range, severe stroke with NIHSS > 25

Note: *Please refer to your institution specific guidelines, these tPA criteria are for instruction and should not to be used directly to guide patient care.*

Case 3 – Stroke

A 54 year old right handed man has high blood pressure, high cholesterol and a history of alcohol abuse. He developed a left sided headache, followed by right arm weakness and right sided sensory loss that quickly spread to involve his right face and leg. On exam his blood pressure is 215/110 mmHg, pulse is 94 and blood glucose is 126. CBC and BMP are unremarkable. He receives 11 points on the NIHSS for mild right facial weakness, complete plegia of the right arm, mild drift of the right leg and dense right sensory loss. However, his language is intact and he has no visual deficits.

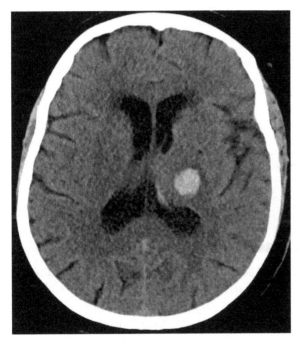

◀ *His non-enhanced head CT is shown here.*

1. What description best describes this patient's head CT?
A. Acute ischemic stroke
B. Acute subarachnoid hemorrhage
C. Chronic basal ganglia calcification
D. Acute thalamic hemorrhage

2. What is the next best step in treatment?
A. Administer IV tPA
B. Admit to the Intensive Care Unit
C. Administer anti-hypertensive medication
D. Administer anti-seizure medications

Case 3 – Diagnosis: Hemorrhagic thalamic stroke

1. What description best describes this patient's head CT?
D. Acute thalamic hemorrhage

2. What is the next best step in treatment?
C. Administer anti-hypertensive medication, blood pressure control and reversing coagulopathies are the mainstay of acute intraparenchymal hemorrhage treatment

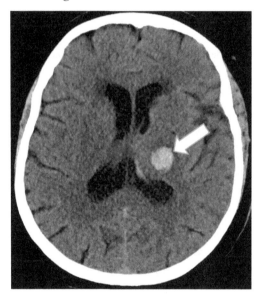

This patient has a hemorrhagic stroke in the left thalamus (arrow). Of the options given, the next best step is to reduce the systolic blood pressure to at least 160 mmHg or less. Lowering systolic blood pressure below 120 mmHg has not been shown to improve outcomes, and may increase complications.

3. What is the most likely cause of the patient's hemorrhagic stroke?
A. Uncontrolled hypertension
B. Hemorrhage into an ischemic stroke
C. Undiagnosed coagulopathy
D. Ruptured intracranial aneurysm

4. Should this patient have a urine drug screen? His spouse denies drug use.
A. Yes, everyone with a cerebral hemorrhage needs urine toxicology
B. Yes, anyone under age 55 needs urine toxicology
C. No, urine toxicology should only be ordered if history suggests drug use
D. No, urine toxicology is not relevant to hemorrhagic stroke

Case 3 – Diagnosis: Hemorrhagic thalamic stroke

3. What is the most likely cause of the patient's hemorrhagic stroke?
A. Uncontrolled hypertension – the thalamus is a typical location

4. Should this patient have a urine drug screen? His spouse denies drug use.
B. Yes, anyone under age 55 needs urine toxicology

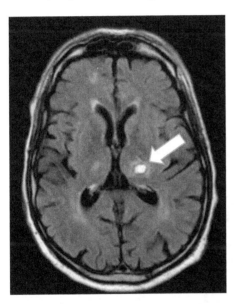

◄ *A T2W FLAIR brain MRI is done 3 weeks later, and continues to show a small area of resolving hemorrhage (arrow). Radiologists typically describe this as 'expected evolution' of the stroke.*

The thalamus is a typical location for a hemorrhagic stroke due to hypertension. Patients will often have severe weakness and sensory loss, but do not typically have language or visual complaints. If blood extends into the ventricles it can impede the outflow of cerebrospinal fluid (CSF) – leading to one of the most feared complications of hemorrhagic stroke, hydrocephalus and death. The treatment for hydrocephalus due to a hemorrhagic stroke is the placement of an external ventricular drain, or EVD. A discussion of EVD's is outside the scope of this text, but by draining CSF they prevent hydrocephalus.

Teaching Point: Urine toxicology is recommended in patients with hemorrhagic strokes anytime there is a clinical suspicion for drug abuse. It is also recommended in anyone age 55 or younger. Stimulants like cocaine or methamphetamine can cause hypertensive hemorrhage!

Case 3 – Hypertensive Hemorrhage

There are several locations that are 'classic' for hypertensive hemorrhage, all located deep within the brain. In rough order of frequency they are:

- Putamen
- Thalamus
- Pons
- Cerebellum

These are all locations where small penetrating arterioles branch directly off larger, higher pressure blood vessels. Chronic high blood pressure leads to weakened vessel walls and eventual rupture.

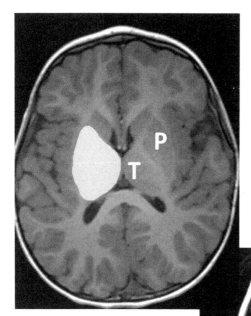

◄ *T1W brain MRI, with common hypertensive hemorrhagic stroke locations indicated by shading, and labeled (P = putamen and T = thalamus).*

▼ *T1W brain MRI also showing typical hypertensive stroke locations as shaded. (Po = Pons, C = Cerebellum)*

Note that hypertensive hemorrhages can occur in other locations – however, these four locations are by far the most common.

Case 4 – Stroke

While shopping, a 52 year old right handed woman suddenly develops left sided weakness, right gaze deviation and confusion. She is rushed to the ER within an hour of symptom onset, where her BP is 180/100 mmHg, pulse is 110 and irregularly irregular, and glucose is 96. She has hypertension and known atrial fibrillation, for which she takes warfarin. On exam she is alert and follows commands, but has a dense left hemiplegia and right gaze deviation. Although sensation on the left is diminished, she can feel the left arm when touched, but has neglect when both arms or legs are touched simultaneously. Her NIHSS is 13. Her labs show an INR of 2.7 and platelet count of 230,000.

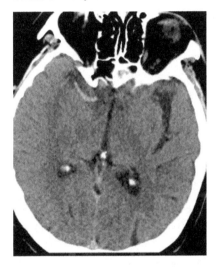

◄ *The patient's non-enhanced head CT. Do you see any suspicious features?*

1. Which vascular territory would you localize the patient's symptoms to?
A. Right middle cerebral artery
B. Left posterior cerebral artery
C. Left anterior cerebral artery
D. Basilar artery

2. Given the patient's symptoms, CT and exam (NIHSS 13), which of the following options is the <u>most likely</u> cause of the stroke?
A. Small vessel disease (lacunar stroke)
B. Stroke mimic
C. Atrial Fibrillation
D. TIA

3. What is the next best therapeutic step in management for this patient?
A. Administer IV tPA
B. Administer labetalol
C. Evaluate for mechanical thrombectomy
D. Admit to the ICU

Case 4 – Stroke

1. To which vascular territory would you localize the patient's symptoms?
A. Right middle cerebral artery – note signs such as gaze deviation and neglect indicate that the cortex is involved, not only deep structures

2. Given the patient's symptoms, CT and exam (NIHSS 13), which of the following options is the <u>most likely</u> cause of the stroke?
C. Atrial Fibrillation – of the options, it is most likely to cause a large vessel occlusion

3. What is the next best therapeutic step in this patient's management?
C. Evaluate for mechanical thrombectomy

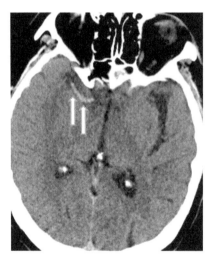

◄ *The patient's symptoms – left hemiplegia, neglect and right gaze deviation, all localize to the right middle cerebral artery territory. This artery supplies the right motor cortex and right parietal lobe. Note the presence of a hyperdensity within the right middle cerebral artery (arrows). This 'hyperdense MCA' is suggestive of a fresh blood clot within the artery.*

The patient's symptoms localize to the right middle cerebral artery territory. Cortical signs such as gaze deviation from involvement of the frontal gaze centers, and neglect to double simultaneous stimulation, suggest that the brain cortex is involved – as opposed to a pure sensory/motor subcortical stroke. The head CT shows no hemorrhage or large infarct, but the artery that we're most concerned about, the right middle cerebral artery, looks bright on non-enhanced CT. This suggests that the artery is filled with a blood clot.

The most common source of a large embolic blood clot, big enough to occlude the middle cerebral artery, is atrial fibrillation. Other causes are possible, including carotid artery atherosclerosis or dissection. This patient had a blood clot despite being on warfarin. Note that being fully anticoagulated, with an INR > 1.7, is a contraindication to IV tPA. Therefore, the patient needs to be evaluated for mechanical thrombectomy – emergently removing the clot from the blood vessel via an endovascular approach.

Case 4 – Large Vessel Occlusion

The patient then undergoes a CT angiogram, which demonstrates an acute cutoff of the right middle cerebral artery, confirming a large vessel occlusion. She is taken to interventional radiology where a catheter is placed in the right femoral artery and advanced to the right internal carotid artery.

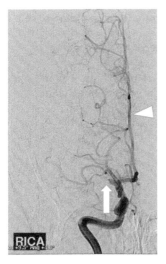

◄ *Catheter angiogram, performed by injecting contrast dye directly into the right internal carotid artery. The ACA is present (arrowhead) but the MCA abruptly cuts off (arrow) due to complete occlusion by a thrombus.*

4. What is the <u>recommended</u> time window for mechanical thrombectomy in an MCA occlusion?
A. 3 hours
B. 4.5 hours
C. 6 hours
D. 12 hours

5. Had the patient's INR been 1.4, rather than 2.7, what treatment should the patient have received?
A. IV tPA only
B. IV tPA, then thrombectomy if no improvement within 30 minutes
C. Mechanical thrombectomy only
D. IV tPA followed by immediate mechanical thrombectomy

The patient had successful removal of the thrombus in the MCA and restoration of blood flow. Interventionalists call successful recanalization and restoration of blood flow a TICI 2b/3 result. TICI stands for *Thrombolysis in Cerebral Infarction* and is pronounced "tikki." TICI 0 indicates complete occlusion.

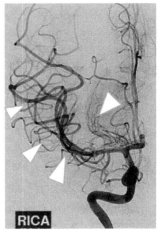

► *Successful recanalization of the MCA (arrowheads).*

Case 4 – Large Vessel Occlusion

4. What is the recommended time window for mechanical thrombectomy in an MCA occlusion?
C. 6 hours from last known well – although note that recent evidence has shown that in carefully selected patients, using perfusion imaging to identify patients with small core infarcts but large areas of brain tissue potentially 'at risk', the time window can be extended up to 24 hours from last known well.

5. Had the patient's INR been 1.4, rather than 2.7, what treatment should the patient have received?
D. IV tPA followed by immediate mechanical thrombectomy – combined therapy, when possible, is the standard of care for large vessel occlusion.

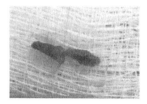

◀ *The successfully retrieved blood clot.*

National guidelines recommend mechanical thrombectomy in the anterior circulation (i.e. the carotid or proximal middle cerebral arteries) within 6 hours of symptom onset. However, patient treatment is often individualized, and advanced imaging techniques such as CT perfusion are often used to select patients for thrombectomy outside the traditional time window. In the posterior circulation (i.e. the basilar artery), the time window is often extended due to the very poor prognosis associated with most brainstem strokes. Thrombectomy confers enormous benefit in reduced mortality and disability, with a number needed to treat to prevent disability of 3.

Patients with an acute stroke due to a large vessel occlusion who are eligible for IV tPA should receive it. They should also be evaluated for thrombectomy – these two treatments are often performed together.

> The DAWN and DEFUSE 3 trials have shown a benefit in using perfusion imaging to select patients who may benefit from thrombectomy in an extended the time window from 6 – 24 hours from last known well. Traditional criteria still apply within the 6 hour window. The WAKE UP trial has shown the possibility of giving IV tPA based on a Diffusion-FLAIR mismatch on MRI. Using imaging to extend the traditional treatment window is a rapidly evolving area of stroke care.

Case 5 – Stroke

A 70 year old right handed man presents to the emergency department after developing significant visual loss in the left eye. One week ago he reports having had two separate episodes of double vision, each lasting ten minutes and which resolved spontaneously. Three days ago he had transient darkening of vision in the left eye, which also resolved spontaneously after about 5 minutes. For the past two months he has had new onset headaches, neck pain and malaise. He reports no significant past medical history and no vascular risk factors. On exam he can see hand motion only in the left eye, vision is normal in the right eye. There is tenderness in the left temple with palpation. A funduscopic exam reveals a pale, swollen optic disc on the left.

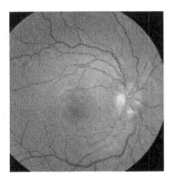

◀ *Ophthalmology reports a pale, swollen optic disc on funduscopic exam.*

1. Which of the following labs are <u>most</u> critical for an elderly patient with temporal tenderness and sudden vision loss, such as this patient?
A. HIV serology
B. ESR/CRP
C. Urine toxicology screening
D. Syphilis serology

2. Which of the following is the next best step in the management of this patient?
A. Treat with high dose IV methylprednisolone
B. Arrange for urgent outpatient ophthalmology the next day
C. Empirically give high dose IV penicillin
D. Arrange for temporal artery biopsy the same day

Case 5 – Giant Cell Arteritis

This patient has Giant Cell Arteritis (GCA). It is one of the most common types of systemic vasculitis, and although it can be systemic, most commonly affects the arteries of the face. Involvement of the ophthalmic artery can lead to blindness, called Arteritic Ischemic Optic Neuropathy (AION). The diagnosis is confirmed with temporal artery biopsy, but this should not delay treatment with high dose steroids. It is a medical emegency as it often becomes bilateral if treatment is delayed.

Clinical Features of GCA
- Affects people age ≥ 50
- New onset headache
- Abrupt vision changes, especially diplopia and monocular vision loss
- New fever, myalgia, malaise and anemia
- Elevated erythrocyte sedimentation rate (ESR) or serum C-reactive protein (CRP)

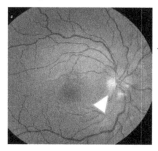

◄ *Swollen optic disc (arrowhead) on funduscopic exam. The optic disc is part of the optic nerve.*

1. Which of the following labs are <u>most</u> critical for an elderly patient with temporal tenderness and sudden vision loss, such as this patient?
B. ESR/CRP strongly support the diagnosis if elevated – although you can treat on clinical suspicion alone

2. Which of the following is the next best step in the management of this patient?
A. Treat with high dose IV methylprednisolone to avoid loss of vision in the other eye

> **Nonarteritic Ischemic Optic Neuropathy (NAION)** Ischemia of the optic nerve can be due to regular old vascular disease as well. Often patients have a history of diabetes, hypertension and tobacco use. Classically, visual loss is precipitated by hypotension and occurs during sleep or after a surgery. Visual loss usually involves either the upper or lower visual fields, a feature called 'altitudinal' visual loss.

Case 5 – Stroke

A 34 year old left handed man awakens with persistent vertigo and nausea, as well as left face pain. He presents to the emergency department where he is found to have loss of pain and temperature sensation on the left face, as well as in the right arm and leg. On finger to nose exam, however, he seems ataxic only on the right side. He seems to have difficulty swallowing, needing to frequently cough to clear his secretions, and he speaks with a slightly hoarse voice. Further history reveals that he had struck the back of his head two days ago while at his job as a construction worker. For the past two days he has had moderate left neck pain, which improved with over the counter naproxen.

1. Given this constellation of symptoms, which other exam finding would you expect to see?
A. Right sided Horner syndrome
B. Left sided Horner syndrome
C. Right sided third nerve palsy
D. Right sided facial nerve palsy

2. What is the most common cause of this syndrome?
A. Basilar artery occlusion
B. Vertebral artery dissection
C. Superior cerebellar artery thrombosis
D. Anterior inferior cerebellar artery dissection

3. Which stroke syndrome does this patient have?
A. Lateral medullary syndrome (Wallenberg syndrome)
B. Superior alternating hemiplegia (Weber syndrome)
C. Paramedian midbrain syndrome (Benedikt syndrome)
D. Medial medullary syndrome (Dejerine syndrome)

Case 5 – Lateral Medullary Stroke

This patient presents with a classic lateral medullary stroke, also known as Wallenberg syndrome. He was found to have a left vertebral artery dissection, and started on antiplatelet therapy. Anticoagulation and antiplatelet medications have been shown to be equally effective for prevention of recurrent stroke after a dissection.

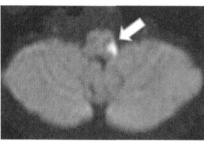

◀ *Diffusion weighted MRI showing a left lateral medullary stroke (arrow).*

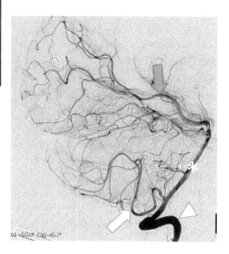

▶ *Conventional catheter angiogram of the posterior circulation, showing the vertebral artery (arrowhead), PICA (white arrow), basilar artery (*) and the posterior cerebral artery (solid arrow). Vertebral artery disease is the most common cause of lateral medullary syndrome, followed by PICA disease.*

Structure	Effect
Vestibular nuclei	Vomiting, vertigo, nystagmus
Inferior cerebellar peduncle	Ipsilateral ataxia, dysmetria
Lateral spinothalamic tract	Loss of contralateral pain and temperature sensation from the body
Spinal trigeminal nuclei (CN V nuclei)	Loss of ipsilateral pain and temperature sensation from the face
Nuclei of CN IX, X	Dysphagia, hoarseness
Descending sympathetics	Ipsilateral Horner syndrome

1. Given this constellation of symptoms, which other exam finding would you expect?
B. Left sided Horner syndrome

2. What is the most common cause of this syndrome?
B. Vertebral artery dissection – the lateral medullary syndrome is usually due to an infarct in the posterior inferior cerebellar (PICA) territory, but the most common cause is a vertebral artery dissection leading to a PICA infarct.

3. Which stroke syndrome does this patient have? **A. Lateral medullary syndrome**

Notes

Paul D. Johnson, MD

Epilepsy

"He's not the Messiah. He's a very naughty boy!"
Life of Brian

A Very Short Introduction to Electroencephalography (EEG)

Much like the heart, or skeletal muscle, the brain has innate electrical activity (electrochemical, actually). In the same way that abnormal electrical activity in the heart can lead to cardiac arrhythmias, abnormal electrical activity in the brain can cause seizures.

The causes of abnormal brain electrical activity can be due to inherited abnormalities of the ion channels in neurons – this genetic cause usually leads to epilepsy presenting in childhood or adolescence. Epilepsy can also be acquired through structural brain injuries causing damage to the brain's "electrical circuitry." A common structural causes in young adults is traumatic brain injury. In the elderly brain tumors or strokes are common. Metabolic or toxic causes can temporarily disrupt brain function enough to cause abnormal electrical patterns leading to provoked, or so called "symptomatic" seizures – these include electrolyte derangements, substance abuse or sepsis.

An electroencephalogram (EEG) records brain electrical activity on the scalp, in a similar manner to measuring cardiac activity with an ECG. However, the electrical signals in the brain are far weaker than the heart's, and the signal is buffered by spinal fluid, bone and scalp. Artifacts from muscle or even electrical equipment in the room can cause artifacts. Therefore, the information that the EEG provides can be messy and challenging to interpret at times.

EEG is performed in real time, meaning it shows current brain electrical activity at the time that the electrodes are on the patient's scalp. If a patient has a seizure just prior to, or just after, the EEG recording, the seizure may be missed and the study could be reported as "normal." An EEG obtained between seizures is called an interictal study.

Although EEG monitoring may pick up overt seizures, they can also detect high risk electrical activity – called "epileptiform discharges." These typically represent chaotic, high amplitude electrical discharges. Focal brain dysfunction, such as focal slowing, can also indicate underlying brain injury and predisposition to seizure.

A typical bedside EEG takes about 30 minutes. Ideally the study is obtained with the patient both awake and asleep. Recording during the transition from wakefulness to sleep helps to detect epileptiform discharges. Longer duration monitoring with video recording, can also be done when the concern for detecting occult seizures is high.

A Very Short Introduction to Electroencephalography (EEG)

Figure 1. An example of a normal EEG recording, note that the EEG has an inherent background frequency or rhythm (see arrow for a good example). An eye blink artifact (arrowhead) is notable here as well.

Figure 2. Example of a generalized epileptic discharge (between arrows). Notice the high amplitude, sharp, rhythmic features.

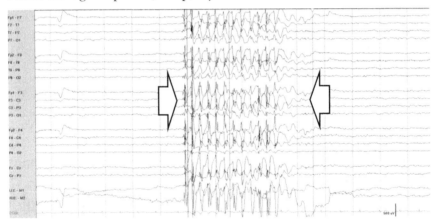

Figure 3. A seizure begins focally (arrowhead) with subsequent spread to other brain areas — note the increased amplitude and rhythmicity over time as the seizure evolves (moving left to right).

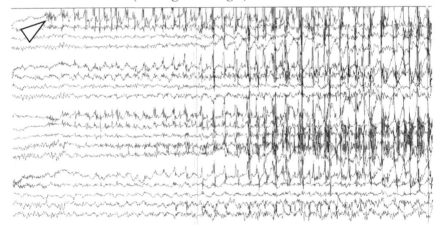

The Classification of Seizures

Seizures are classified both from where the seizure arises within the brain, and by its clinical effects. The International League Against Epilepsy (ILAE) publishes up-to-date seizure classifications on its website, www.epilepsy.com.

I. Focal Onset The seizure arises from a discreet location within the brain, from a specific hemisphere, and usually from a specific brain lesion (i.e. a stroke, tumor, head injury, congenital malformation, etc.) Clinically this would be any seizure with identifiable focal onset.
    A. Aware/Unaware Take note of the patient's level of awareness during the event. They can be aware even if unable to speak – but awareness is impaired if there is any degree of confusion.
    B. Motor/Non-motor onset Is there a motor component, i.e. twitching or convulsions? Non-motor symptoms include changes in thinking, altered sensation, etc.

II. Generalized Onset The seizure involves both cerebral hemispheres at the onset, often due to a genetic condition, such as abnormal neuronal $Na+/K+$ channels, or a systemic effect like alcohol withdrawal. Often accompanied by loss of consciousness.
    A. Motor vs Non-motor Was the seizure convulsive or non-motor (i.e. an absence seizure, or staring spell).

III. Unknown onset If you can't clearly tell if the seizure had a focal onset, for example if the onset was not observed, it is OK to call it a seizure of unknown onset type.

Other Terms

Aura: An aura is the symptoms a patient feels at the beginning of a seizure. It is actually the start of a focal seizure.

Focal to generalized seizures: Many seizures will have a focal start, but then become generalized as the abnormal electrical activity spreads across the brain, often referred to as 'secondary generalization.'

Epilepsy: Defined as having two or more recurrent, unprovoked seizures > 24 hours apart. Note that a single lifetime seizure is not sufficient to diagnose epilepsy. Provoked seizures, for example alcohol withdrawal seizures, do not count towards the diagnosis of epilepsy.

The description of the onset and features of the seizure are called the **seizure semiology.**

Introduction to Antiepileptic Medications

Drug	Notes	Potential Adverse Effects
Phenobarbital*	Long half-life, mean 80 hours	Sedation, ataxia and dysarthria at toxic levels
Phenytoin*† (Dilantin)	Can monitor serum levels, the free form is the active form	Nystagmus, ataxia. With chronic use hirsutism, gingival hyperplasia. IV form can cause bradycardia & phlebitis
Carbamazepine (Tegretol)	Can also be used to treat trigeminal neuralgia	Nausea, dizziness, weight gain, Stevens-Johnson syndrome (higher risk in SE Asians), hyponatremia
Valproate* (Depakote)	Best for juvenile myoclonic epilepsy	Tremor, hair loss, weight gain, hepatotoxicity, hyper-ammonemia
Gabapentin	Rarely used except in elderly	Poor efficacy, fatigue, weight gain, good for nerve pain
Lamotrigine (Lamictal)	Needs a very slow titration, watch for rash	Stevens-Johnson syndrome (high risk with rapid titration or if combined with valproate)
Topiramate (Topamax)	Good for migraines at 50mg BID	Weight loss, paresthesias, renal calculi, rare hepatic failure, cognitive slowing
Levetiracetam* (Keppra)	Very few drug interactions, renally cleared	Agitation, psychosis
Zonisamide (Zonegran)		Stephens-Johnson syndrome, renal calculi
Lacosamide* (Vimpat)	Very few drug interactions, renally cleared	Diplopia, rarely causes heart block
Pregabalin	Poorer efficacy	Weight gain, pedal edema

* Can be loaded IV; † can use IV fosphenytoin for fewer side effects

Note: The combination of valproate and lamotrigine requires dose adjustments, as valproate significantly increases lamotrigine levels. Likewise, valproate, phenytoin and carbamazepine all affect warfarin metabolism.

Carbamazepine and phenytoin share a similar mechanism of action, and may cause more adverse effects when used together.

Case 1 – Epilepsy

A 21 year old man is brought to the emergency department by ambulance after he suffered his first generalized tonic-clonic seizure. The seizure lasted one minute, and the man has been slowly waking up but remains slightly confused. He denies any history of head trauma, prior meningitis, alcohol or substance abuse. In fact, he is a college student, who has recently been cramming for his mid-term exams. He denies any prior medical problems or family history of seizures.

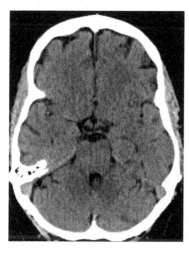

◄ *His non-enhanced head CT was read by the radiologist as showing "No acute intracranial abnormality."*

1. After a single, unprovoked seizure, should the patient be diagnosed with epilepsy?
A. Yes   B. No   C. Only if they have a family history of seizure

2. After a single, unexplained seizure, should patients be routinely started on an antiepileptic medication?
A. Yes   B. No   C. Only if they have a family history of seizure

3. Is urine toxicology indicated for routine evaluation of first time seizure?
A. Yes, always   B. No, not routinely

4. Is head imaging indicated after a first seizure?
A. Yes   B. No   C. There are no guidelines

Case 1 – First Seizure

This young man presents with a first ever, unprovoked, generalized tonic-clonic seizure and is returning to his baseline mental status. He had no red flags – no fever or other evidence to suggest meningitis, no known immunocompromised state, no evidence of subarachnoid hemorrhage, and no recent head trauma. In that case, once fully recovered, he can likely be evaluated on an outpatient basis with EEG and contrast enhanced MRI.

Epilepsy is defined as having a predisposition to unprovoked seizures. People are diagnosed with epilepsy after having <u>two</u> unprovoked seizures at least 24 hours apart, or a single unprovoked seizure and a high risk circumstance for recurrent seizure. High risk features include preexisting brain injuries such as stroke or severe traumatic brain injury.

Antiepileptic medications are typically not started after a first seizure – many people have a single seizure without going on to have epilepsy. After a second unprovoked seizure the risk of additional seizures is high, and antiepileptic medications are typically started. I often start phenytoin in the ER.

Urine toxicology is only recommended when a clinical suspicion for substance abuse exists. All patients should have head imaging after a seizure. A brain MRI with contrast and an EEG are typically done as well, often on an outpatient basis.

1. After a single, unprovoked seizure, should the patient be diagnosed with epilepsy?
B. No – epilepsy diagnosis requires two or more unprovoked seizures

2. After a single, unexplained seizure, should patients be routinely started on an antiepileptic medication?
B. No – because of the many side effects of treatment, and many patients with a single unprovoked seizure will not go on to have additional seizures.

3. Is urine toxicology indicated for routine evaluation of first time seizure?
B. No, not routinely unless a specific concern for substance abuse exists

4. Is head imaging indicated after a first seizure?
A. Yes – ideally a brain MRI with and without contrast to look for a seizure focus

> **Teaching Point:** Patients with seizures should always be given seizure precautions – recommendations on how to avoid injury if another seizure occurs. This includes avoiding swimming, bathing alone or working at heights. Driving laws after a seizure are variable from state to state.

Case 2 – Epilepsy

A 28 year old woman has known temporal lobe epilepsy, which has been well controlled on low dose levetiracetam (Keppra), with very few breakthrough seizures in the past few years. When untreated she is prone to frequent seizures, some of which are convulsive seizures. She calls the clinic to report that she has recently discovered that she is 10 weeks pregnant.

1. What is the next best step in management?
A. Hold levetiracetam until the 3rd trimester
B. Hold levetiracetam until the 2nd trimester
C. Hold levetiracetam for the duration of pregnancy
D. Continue levetiracetam

2. Which of the following antiepileptic medication has the <u>highest</u> risk of congenital malformation when used during pregnancy?
A. Valproic acid
B. Carbamazepine
C. Phenytoin
D. Lamotrigine

3. Women with epilepsy who are planning to conceive should undergo which of the following interventions?
A. Vitamin K supplementation
B. Folate supplementation
C. Thiamine supplementation
D. Long term video EEG monitoring to assess seizure control

4. Which of the following antiepileptic medications is most readily passed through breast-milk?
A. Phenobarbital
B. Phenytoin
C. Carbamazepine
D. Levetiracetam

> **Breastfeeding with Epilepsy:** Note that some antiepileptics do cross into breast milk, especially barbiturates (i.e. phenobarbital), benzodiazepines, ethosuxamide and lamotrigine. Infants usually tolerate this, but should be monitored for side effects.

Case 2 – Epilepsy in Pregnancy

Many antiepileptic medications are implicated in causing congenital malformations – the risks associated with these medicines are actually well known. In general antiepileptic medications are continued for women with a high risk of convulsive seizures, as convulsions carry a high risk of both maternal and fetal injury. In addition, all women trying to conceive should be on folate supplementation to reduce the risk of birth defects. Most women do not experience a change in seizure frequency during pregnancy.

1. What is the next best step in management?
D. Continue levetiracetam

2. Which of the following antiepileptic medication has the highest risk of congenital malformation when used during pregnancy?
A. Valproic acid

3. Women with epilepsy who are planning to conceive should undergo which of the following interventions?
B. Folate supplementation

4. Which of the following antiepileptic medications is most readily passed through breast-milk?
A. Phenobarbital

Highest Risk Antiepileptics during Pregnancy:

<u>Valproic Acid</u>: Highest risk medication, causes midline defects such as spina bifida in an estimated 6-9% of exposed children

<u>Phenobarbital</u>: Cardiac defects, passed through breast milk

<u>Topiramate</u>: Causes oral clefts

Overall risk of birth defects in mothers with epilepsy: 4-6%

Probably the Safest Antiepileptics:

Lamotrigine
Levetiracetam
Phenytoin
Carbamazepine

Further Reading?

Check out: Harden CL Pregnancy and Epilepsy. *Continuum.* AAN, 2014.

Tip: Certain antiepileptics, such as carbamazepine and phenytoin, can increase the metabolism of birth control – leading to contraceptive failure and unintended pregnancy! Also be aware that pregnancy accelerates the metabolism of lamotrigine, and doses may need to be temporarily increased.

Case 3 – Epilepsy

A 71 year old right handed man was on warfarin for atrial fibrillation. He fell three days prior to admission and developed a right sided subdural hematoma, with significant midline shift and resulting mental status changes. He went to the OR where a hemicraniectomy was performed and the subdural blood was drained. The skull has been replaced (cranioplasty) – he is now in the ICU but without good recovery of his cognitive functions. He has been noted to have some left face twitching, and an EEG was ordered.

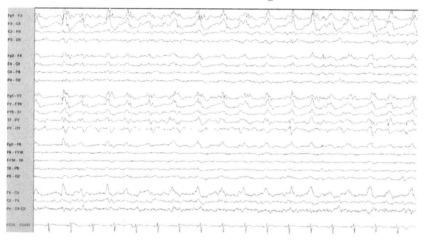

▲ *The EEG is read as showing "Periodic Lateralized Epileptiform Discharges" or PLEDs.*

1. Does the presence of epileptiform discharges mean the patient is currently having seizures?
A. Yes   B. No   C. Sometimes

2. There is either no risk or low risk of having seizures when periodic epileptiform discharges are present.
A. True   B. False

3. Lateralized epileptiform discharges typically indicate an underlying focal brain injury.
A. True   B. False

4. Antiepileptic drug treatment should always be increased until PLEDs disappear on the EEG.
A. True   B. False

Case 3 – Periodic Epileptiform Discharges

Periodic lateralized epileptiform discharges (PLEDs) are relatively common in very ill patients with focal brain injuries – it is especially common in the neurosurgical ICU. They represent injured brain tissue with a high risk for seizures, but are not seizures themselves. Treating them can very difficult – as they may not disappear on EEG without significant sedation. PLEDs by themselves are not necessarily harmful to patients. Most medical providers will treat with antiepileptic medications with the goal of preventing overt seizures while balancing the sedating side effects of medication. Often PLEDs resolve on their own within several days or weeks.

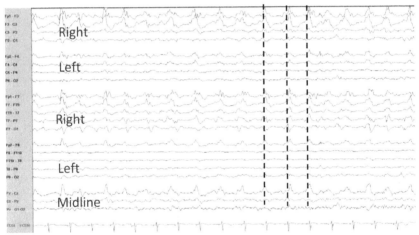

▲ *This EEG shows epileptiform discharges (a few are indicated by dotted lines) every second or so (i.e. periodic), all arising from the same location (lateralized), indicating an underlying brain injury with high risk features for focal seizure. Note that channels correspond to the right, left or midline brain cortex (labeled). You can see that the PLEDs arise from the injured right hemisphere.*

1. Does the presence of epileptiform discharges mean the patient is currently having seizures?
B. No – but they do indicate irritated brain from which a seizure could arise

2. There is either no risk or low risk of having seizures when periodic epileptiform discharges are present.
B. False – epileptiform discharges represent high risk for seizure

3. Lateralized epileptiform discharges typically indicate an underlying focal brain injury.
A. True – the discharges arise from a discreet area of injured brain

4. Antiepileptic drug treatment should always be increased until PLEDs disappear on the EEG. **B. False – treatment should be tailored to the patient; PLEDs can be difficult to stop completely without resorting to general anesthetics, which may not always be in the patients best interest**

Antiepileptic Medication Review

Match the antiepileptic medication listed below to the correct side effect. Each one has a single best answer and none are repeated.

- A. Levetiracetam
- B. Carbamazepine
- C. Valproic acid
- D. Pregabalin
- E. Topiramate
- F. Lamotrigine
- G. Phenytoin

1. Half-life is significantly increased when combined with valproic acid, risk of Stephens-Johnson syndrome

2. May cause depression and irritability

3. Can cause SIADH and hyponatremia, especially in the elderly

4. Causes paresthesias and calcium phosphate kidney stones

5. May cause phlebitis when given intravenously due to its very alkaline pH

6. Causes weight gain and pedal edema

7. May cause alopecia, obesity and tremors

Drug Levels: The utility of checking antiepileptic drug levels various from medication to medication. Phenytoin has the most useful drug levels, with a therapeutic range generally between 1.5 – 2. Above 2.5 patients may develop dizziness and nystagmus. Checking levels in most other medications, such as levetiracetam, is not very useful unless you want to see if the patient is not taking the medication at all, in which case an undetectable level is helpful.

Antiepileptic Medication Review Answers

Match the antiepileptic medication listed below to the correct side effect. Each one has a single best answer and none are repeated.

   A. Levetiracetam
   B. Carbamazepine
   C. Valproic acid
   D. Pregabalin
   E. Topiramate
   F. Lamotrigine
   G. Phenytoin

1. Half-life is significantly increased when combined with valproic acid, risk of Stephens-Johnson syndrome
   F – lamotrigine

2. May cause depression and irritability
   A – levetiracetam

3. Can cause SIADH and hyponatremia, especially in the elderly
   B – carbamazepine

4. Causes paresthesias and calcium phosphate kidney stones
   E – topiramate (can also induce acute angle glaucoma)

5. May cause a phlebitis when given intravenously due to its very alkaline pH
   G – phenytoin (the 'purple glove' syndrome)

6. Causes weight gain and pedal edema
   D – pregabalin, note gabapentin can do this as well

7. May cause alopecia, obesity and tremors
   C – valproic acid

Case 4 – Epilepsy

A 31 year old woman is brought to the emergency department from an outpatient clinic. She was about to leave the office, having just finished a routine appointment and blood draw, when she became pale, stared into the distance for several seconds and then collapsed with loss of consciousness. Witnesses describe several seconds of generalized, tonic-clonic appearing shaking of the arms and legs. The woman also bit her tongue. She awoke and returned to normal over five minutes. She then describes having felt nauseated just prior to the loss of consciousness. There is no history of prior similar events.

1. What is the most likely diagnosis?
A. Provoked tonic-clonic seizure
B. New onset adult epilepsy
C. Vasovagal syncope
D. Focal seizure with secondary generalization

2. Which study needs to be done for a seizure work-up in this patient?
A. Head CT
B. Brain MRI
C. EEG
D. All of the above
E. None of the above

3. All of the following can be seen in vasovagal syncope, except which?
A. Prolonged time to return to consciousness (> 1 hour)
B. Tongue biting
C. Urinary incontinence
D. Convulsive shaking

Case 4 – Physiologic, Non-Epileptic Spells

This patient had a classic example of vasovagal syncope, in which high parasympathetic tone results in low heart rate and brief loss of consciousness. Classic triggers include prolonged standing, blood draws or using the bathroom (micturition syncope). The loss of consciousness is usually brief, with patients awakening and returning to baseline within minutes of the event. They often appear pale or ashen, feel nauseated or flushed, and may have tunnel vision prior to the event. Tonic-clonic jerking of the extremities can occur when brain mediated control over the spinal cord is interrupted. Tongue biting and urinary incontinence are uncommon, but can occur during syncope. Seizure is a clinical diagnosis, and when the history is entirely consistent with vasovagal syncope, as in this case, no further diagnostic studies are indicated.

Another cause for loss of consciousness can be cardiac arrhythmia, and in the appropriate clinical context additional cardiac evaluation, such as an EKG or prolonged cardiac telemetry, may be indicated.

1. What is the most likely diagnosis?
C. Vasovagal syncope – she has many classic features

2. Which study needs to be done for a seizure work-up in this patient?
E. None of the above – seizure is a clinically diagnosis, and clinically she had vasovagal syncope

3. All of the following are seen in vasovagal syncope, except which?
A. Prolonged time to return to consciousness (> 1 hour), all the rest can be seen in vasovagal syncope at times

> **Psychogenic, non-epileptic spells:** Although patients may have a physiologic, non-epileptic (i.e. not seizure related) reason for loss of consciousness, they may also have a psychogenic non-epileptic spell. Characteristics of psychogenic, non-epileptic events includes arrhythmic tonic-clonic movements, provocation by anxiety, history of PTSD and keeping the eyes shut during events (eyes are often open during epileptic seizures). In these cases capturing an event on EEG may be necessary to exclude epilepsy. Patients commonly have both true epileptic and psychogenic events.

Case 5 – Epilepsy

A 71 year old left handed man with diabetes, prior TIA and hypertension presents to the emergency department after his first seizure, which started with arm stiffening and moaning, followed by a generalized convulsion. The event lasted 45 seconds and stopped spontaneously. He has no prior personal or family history of seizure. Two years ago he had a severe traumatic brain injury with loss of consciousness, but he has recovered well and is living at home. The head CT was read as "negative."

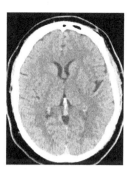

1. What is the most common cause of new onset epilepsy in this patient's age group?
A. Medication side effects
B. Brain tumors
C. Traumatic brain injuries
D. Cerebrovascular disease

2. For a patient with a single, spontaneous seizure, which of the following is not a high risk feature for recurrent seizures?
A. Sharp spikes on EEG
B. Seizure occurrence during sleep
C. History of traumatic brain injury
D. The first seizure is a generalized convulsion

3. Having a seizure within what time frame following a traumatic brain injury (TBI) carries the highest risk for new epilepsy?
A. Immediate (within 24 hours of TBI)
B. Within 7 days of the TBI
C. A week or more after the TBI

4. What should this patient be told about his ability to drive following this seizure?
A. No driving for 1 month
B. No driving for 3 months
C. No driving for 6 months
D. Driving restrictions vary by state

5. Should you recommend starting this patient on antiepileptic medications?
A. Yes   B. No

Case 5 – Causes of Epilepsy in Adults

This patient most likely had a delayed seizure from his prior, severe traumatic brain injury (TBI). Patients who have new onset of seizures more than one week from their TBI have a high risk of epilepsy – therefore he should be started on an antiepileptic medication. Seizures occasionally occur immediately at the time of a TBI, and although these are often treated with antiepileptic medications as well, the risk for developing epilepsy is actually lower than in those who develop late seizures.

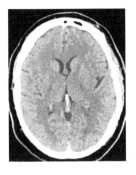

1. What is the most common cause of epilepsy in this patient's age group?
D. Cerebrovascular disease – any focal brain injury can lead to epilepsy, stroke is a most common cause of brain injury in adults

2. For a patient with a single, spontaneous seizure, which of the following is <u>not</u> a high risk feature for recurrent seizures?
D. The 1st seizure is a GTC seizure – the other options are high risk

3. Having a seizure within what time frame following a traumatic brain injury (TBI) carries the highest risk for new epilepsy?
C. A week or more after the TBI – delayed onset is more worrisome

4. What should this patient be told about his ability to drive following this seizure?
D. Driving restrictions vary by state

5. Should you recommend starting this patient on antiepileptic medications?
A. Yes – a history of severe TBI is high risk for recurrent seizures, so in his case starting a medication should be considered. Given his age and history of TIA, a small cortical stroke could also be the culprit

---

Common Causes of First Seizure by Age

- ✓ 2-3rd decade: Trauma (note that the onset of epilepsy after TBI may be delayed by years)
- ✓ 4-5th decade: Brain tumor
- ✓ Elderly: Cerebrovascular disease
- ✓ Note: Neonatal seizures, childhood febrile seizures, and provoked seizures (i.e. those due to acute illness, medication or alcohol withdrawal, etc.) do not count towards a diagnosis of adult epilepsy

Case 6 – Epilepsy

A 52 year old man with known epilepsy is brought to the emergency department by ambulance following a prolonged generalized tonic-clonic seizure. He received 4mg of IV lorazepam in the ambulance and his seizure stopped but he has been somnolent and unresponsive since that time, about 20 minutes in total. While you are examining him you notice that he has a strong left gaze deviation that cannot be overcome. He then develops rhythmic left face and arm twitching. He is prescribed levetiracetam as a home medication but is reportedly non-compliant.

1. How long must a seizure last to qualify as status epilepticus?
A. 5 minutes
B. 10 minutes
C. 15 minutes
D. 30 minutes

2. What is the generally accepted first line antiepileptic medication in status epilepticus?
A. Benzodiazepines
B. Topiramate
C. Levetiracetam
D. Carbamazepine

3. For patients with persistent altered mental status lasting longer than 30 minutes after a generalized seizure, which study would be most helpful?
A. Outpatient brain MRI with contrast
B. Urine toxicology
C. Continuous EEG monitoring
D. 30 minute bedside EEG

4. Which of the following medications would be most appropriate therapy for a patient in status epilepticus who failed to respond to benzodiazepines?
A. Topiramate 300 mg PO
B. IV Lacosamide 50 mg
C. IV fosphenytoin 20 PE/kg
D. Lamotrigine 200 mg PO

Case 6 – Status Epilepticus

Status epilepticus is defined as any generalized seizure lasting for 5 minutes or longer, and also includes two or more seizures if there is no return of consciousness between them. In this patient's case, he was likely in subclinical (i.e. non-convulsive) status between his two seizures. Prolonged status epilepticus is a medical emergency, and can cause permanent brain injury.

When the patient fails to awaken after a seizure and you suspect subclinical status epilepticus, continuous bedside EEG monitoring is indicated. It may take 24 hours of monitoring or more to detect subclinical seizures.

There are many protocols for treatment of status epilepticus. After addressing airway, breathing and circulation, first line medical treatment is with benzodiazepines, usually lorazepam 2-4mg repeated if necessary. Second line treatment is often IV fosphenytoin 18-20 mg/kg or IV valproic acid 40 mg/kg. Third line therapy usually requires intubation and sedative use.

Evaluation for reversible causes of status epilepticus, such as new stroke, hypoglycemia or infection is necessary for refractory status epilepticus. Brain imaging, antiepileptic drug levels, blood count, metabolic panel and renal function are often ordered. Many providers give empiric dextrose with thiamine as well.

1. How long must a seizure last to qualify as status epilepticus?
A. 5 minutes of continuous seizure, or 5 minutes between two seizures without a return of consciousness in between

2. What is the generally accepted first line antiepileptic medication in status epilepticus? **A. Benzodiazepines – up to 0.1 mg/kg up to 10 mg total in divided doses**

3. A patient with persistent altered mental status, lasting longer than 30 minutes after a generalized seizure, should undergo which study?
C. Continuous EEG monitoring to assess for subclinical seizures

4. Which of the following medications would be most appropriate therapy for a patient in status epilepticus who failed to respond to benzodiazepines?
C. IV fosphenytoin at 20 PE/kg (*note the dosing; PE is phenytoin equivalents*). Fosphenytoin has less side effects than IV phenytoin.

For further reading see: Hirsch LJ, Gaspard N. Status Epilepticus. *Continuum.* 2013; 19(3):767-794.

Treatment Algorithm for Status Epilepticus

Here is a proposed algorithm for treating convulsive status epilepticus. Recall that status is defined as a seizure lasting 5 minutes or greater, or back to back seizures without recovery in between. Status is an emergency not only because of compromised airway and potential injury to the patient, but prolonged seizures can cause permanent brain injury.

Status Epilepticus: Initial 5 Minutes

- ABC's (airway, breathing, circulation)
- Fingerstick glucose, thiamine & 50 ml of D50W if low/unknown glucose
- IV lorazepam 2-4 mg push, repeat x 1 if seizing in 5 minutes
- Labs: CBC, BMP, Ca, Mg, PO4, troponins, LFTs, ABG, consider tox screen, HCG (for women), AED levels if known to be on AED's

If status continues, management in next 30 minutes

- IV antiepileptic load. Choose one, but can be combined:
    - Fosphenytoin: 20 PE/kg, can rebolus 5 PE/kg*
    - Phenytoin: 20 mg/kg, can rebolus 5 mg/kg
    - Valproate: 40 mg/kg IV, rebolus at 20 mg/kg if needed
- Alternatively, consider rapid sequence intubation and sedation with propofol or midazolam infusion

If status continues, management after 30 minutes

- Intubated and sedation if not already done
- Consider pentobarbital infusion for
    - Load 5 mg/kg at 50 mg/min until seizure stops
- Continuous EEG monitoring if not already
- Consider additional brain imaging, LP for work-up of refractory status

* Recall that fosphenytoin has a better safety profile than IV phenytoin.

Treatments for Medically Refractory Epilepsy

The Ketogenic Diet is a special high protein, low fat diet, which creates a slightly acidic pH in the body. This can be very useful in difficult to control seizures, but is only practical in children who can't "choose what they eat" as it's almost impossible to comply with otherwise!

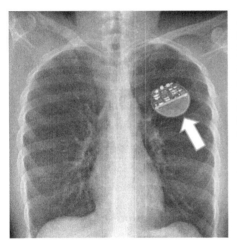

Another option for medically refractory epilepsy is a **Vagal Nerve Stimulator** (VNS). These surgically implanted devices usually add about as much seizure protection as a single antiepileptic medication. The device can be activated with a magnet – which also means they will need to be reset after an MRI scan.

◄ *Vagal nerve stimulator (arrow) implanted under the skin of the left breast.*

Epilepsy surgery can be performed to remove a part of the brain that is causing focal onset seizures and is refractory to medical therapy. Significant care must be taken to identify the exact location of seizure onset – often this includes implanting intracranial electrodes for EEG monitoring, not an easy task! Surgery most often includes the removal of a temporal lobe, but other lesions can be removed as well. Long term success rates are still about 50%.

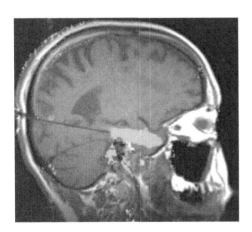

◄ *Hippocampal laser ablation is an option for treating focal temporal lobe seizures - 80% of temporal lobe seizures arise from the hippocampus. This is a relatively minimally invasive surgical procedure, but patients must undergo extensive screening to ensure they are good candidates and won't suffer memory or language dysfunction after the procedure.*

Pediatric Epilepsy Syndromes

There are a number of pediatric epilepsy syndromes which the adult neurology provider need to be familiar with, but will not see frequently. A brief overview is provided here. These genetic syndromes are referred to as primary epilepsies.

Absence Seizures typically present in childhood with short staring spells. Children often appear to be awake and minimally interactive. These can be very brief and go unnoticed for months or years before being detected. They can often be provoked with hyperventilation. Children may grow out of this seizure type. This seizure type responds very well to ethosuximide.

Juvenile Myoclonic Epilepsy a triad of seizure types, include absence seizures (staring spells), early morning myoclonic jerks and generalized tonic-clonic seizures. Patients often present in teens or late 20's and so may be seen by adult neurologists. Phenytoin may actually makes the seizures worse – but they respond extremely well to valproic acid.

Rasmussen Encephalitis is a progressive seizure condition in children, characterized by progressive brain atrophy. The treatment is surgical removal of the affected brain hemisphere (hemispherectomy).

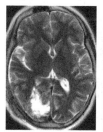
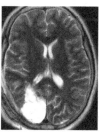
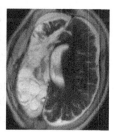

◀ *T2W MRI showing progressive brain atrophy (left to right) in Rasmussen Encephalitis.*

Benign Rolandic Epilepsy (also known as Benign Childhood Epilepsy with Centrotemporal Spikes) represents almost one quarter of all childhood onset epilepsies. The majority of cases present as nocturnal motor seizures in children between 9 – 10 years of age, and usually resolves within 5 years of presentation. This genetic condition responds very well to carbamazepine, although not all children require treatment.

Pediatric Epilepsy Syndromes

Febrile Seizures Between 3 – 5% of all children have a febrile seizure. Age of onset is between 5 months and 5 years, with 90% of febrile seizures occurring before 3 years of age. Febrile seizures that last for 15 minutes or less are considered simple febrile seizures. Less than 5% of children with simple febrile seizures will develop epilepsy as an adult.

Complex febrile seizures are those that last longer than 15 minutes, occur multiple times within a 24 hour period, or have focal onset. Complex seizures carry a worse prognosis, with increased risk of developing epilepsy.

Lennox-Gastaut Syndrome is a childhood epileptic encephalopathy syndrome, meaning that in addition to seizures the child typically has severe cognitive impairment as well. There are many seizure types, which can be difficult to control. The prognosis is generally poor.

Dravet Syndrome is myoclonic epilepsy of childhood, usually presenting within the first year of life. It progresses to have multiple seizure types, and the prognosis is grave.

Tips for interpreting EEG reports

- Most EEG's are performed in between seizures, and therefore are called 'interictal' studies (the ictus being the seizure)
- A typical EEG lasts 30 minutes and captures wakefulness and sleep
- You can request continuous monitoring for inpatients, which provides many hours to days of EEG, usually reserved for patients who may be having non-convulsive (subclinical) seizures
- The skull insulates the brain's electric signals - any break in the skull (i.e. from prior neurosurgery) will cause a **'breach artifact'**
- Any movement, even eye movement, can cause EEG artifact
- **Epileptiform discharges** represent 'irritable' brain tissue and are a high risk feature for seizures
- **Focal slowing** represents dysfunctional brain tissue, and can indicate a higher risk for focal seizures (but can also just show focal injury or dysfunction of any cause)
- Generalized slowing correlates with encephalopathy
- Finding **triphasic waves** suggest a metabolic encephalopathy, like hyperammonemia, or uremia
- **Posterior dominant rhythm (PDR)** normal finding consistent with a functioning, non-impaired brain
- **PLEDs** high risk feature associated with ongoing focal seizures

Case 7 – Epilepsy

A 24 year old woman with epilepsy has frequent focal seizures with alteration of awareness (also called complex partial seizures). During these events, her left arm stiffens and her right hand picks aimlessly at her clothing. An MRI was obtained as part of her epilepsy evaluation, and is shown below.

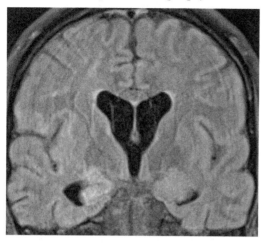

◄ *T2W FLAIR brain MRI. The coronal view is preferred for evaluating the temporal lobes, a common location for focal onset seizure as the hippocampus can be prone to seizures if injured.*

1. From which part of the brain do the seizures likely originate?
A. Right temporal
B. Left temporal
C. Right frontal
D. Left frontal

2. What feature would you <u>not</u> expect to see during this patient's seizure?
A. Automatisms
B. Abnormal smell
C. Mirthless laughter
D. Confusion

3. A patient with temporal lobe epilepsy who has failed two adequate seizure medications is how likely to respond to a third medication?
A. 5-10%
B. 10-20%
C. 20-30%
D. 30-50%

Case 7 – Temporal Lobe Epilepsy

This patient has typical clinical findings of temporal lobe epilepsy, including episodes of confusion with orofacial automatisms (i.e. lip smacking, picking and posturing with the hands). Since the seizure comes from the right temporal lobe, in this case, the left hand is stiff, while the left hemisphere is free to make the right hand pick aimlessly. Other features include an olfactory aura.

Temporal lobe epilepsy that fails multiple antiepileptic drugs can be considered medically refractory. One treatment option for medically refractory temporal lobe epilepsy includes surgical removal of the affected lobe – although seizure control is not always complete and removal of the dominant lobe can have serious cognitive side effects.

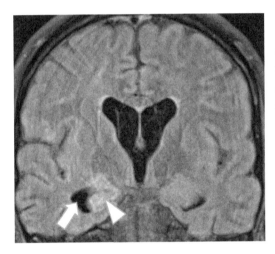

◀ *Notice the right temporal lobe atrophy (arrow), a common finding in temporal lobe epilepsy. The hippocampus is readily apparent (arrowhead).*

1. From which part of the brain do the seizures likely originate?
A. Right temporal – the seizing right brain causes left arm stiffness, the non-seizing left brain causes the right arm to move aimlessly

2. What feature would you <u>not</u> expect to see during this patient's seizure?
C. Mirthless laughter (laughter can be a feature of seizures, typically called 'gelastic seizures', due to hypothalamic lesions)

3. A patient with temporal lobe epilepsy who has failed two adequate seizure medications is how likely to respond to a third medication?
A. 5-10% - patients who fail two medications at appropriate doses may have 'resistant' epilepsy which is hard to control medically

Notes

Paul D. Johnson, MD

Headache

"Nobody calls me Lebowski. You got the wrong guy. I'm the Dude, man."
The Big Lebowski

Headache Basics

Headache is one of the most common reasons why people see a doctor, including both neurology and primary care. More than 10% of adults have migraine headaches and 1% of the population has chronic migraines, which are defined has a migraine heading occurring 15 or more days per month (only 8 of which must have 'typical' migraine symptoms) for at least 3 months. Not surprisingly, headache is also one of the most common causes of disability worldwide.

Headaches can be **primary** – i.e. an idiopathic headache condition such as migraine, tension type headache or the trigeminal autonomic cephalgias (TAC's) or they can be **secondary**, or acquired due to some other condition such as trauma, infection, aneurysm or dissection, medication or tumor, etc. Differentiating between them is a major task for the neurologist, as each requires different evaluation and treatment.

Migraine headaches can be due to a multitude of factors, including genetics, environment, hormones, exacerbating medications and diet. Due to the genetic component, migraines are often familial. Due to the hormonal component they may be associated with menses, called catamenial migraine. They also tend to improve after menopause. Identifying provoking factors, if applicable, and avoiding them is important.

Migraines begin with altered brain connectivity, leading to brainstem activation and abnormal cortical depolarization – all of which contributes to migraine auras, nausea, vomiting, and the very common complaint of migraine associated vertigo.

> **Migraine and Stroke:** Patients with migraine headaches aura have an elevated risk of thromboembolic strokes, especially if they also smoke cigarettes or use medications that cause hypercoagulability, such as estrogen containing oral contraceptives. These patients also have higher rates of PFO than the rest of the population, although the link between this and stroke risk is not clear. Lastly, patients with a history of stroke or coronary artery disease should avoid using triptans and DHE, as these medications could theoretically cause vasospasm and precipitate ischemia.

Tension Type Headache

The most common type of headache is the Tension Type Headache (TTH), which despite its name is *not* related to either musculoskeletal tension or anxiety. TTH is a rather bland type of headache.

Characteristic features of TTH include:
- Mild to moderate pain; unlike migraines, patients can often 'work through' the discomfort and it is not aggravated by physical activity in the way that migraines are
- Duration is generally less than 4 hours, as opposed to migraines which almost always last longer than 4 hours
- Often bilateral
- Throbbing, dull or band-like
- Not usually associated with nausea, vomiting or vertigo

Acute Treatment can include over the counter analgesics, especially NSAIDS such as ibuprofen 200-400 mg, naproxen sodium 220-550 mg or aspirin 650-1000 mg. The addition of caffeine 64-200 mg may be helpful. The chronic use of analgesics or opioids should be avoided (see section on medication overuse headaches later in this chapter).

Treatment of Chronic TTH may include use of tricyclic antidepressants such as nortriptyline or amitriptyline. Always start at the lowest dose and increase slowly. It can take months to see benefit from preventive headache medications. Other options include venlafaxine, gabapentin and topiramate.

Research suggests that just as important as medications are behavioral interventions, especially the following:
- Regular sleep, exercise and meals
- Relaxation
- Biofeedback techniques (a form of mindfulness)
- Some providers try acupuncture or physical therapy

Prophylactic treatment is recommended if the headaches are common, disabling and long lasting.

Need additional resources for headache patients? Consider recommending "Heal Your Headache: The 1-2-3 Program" by David Buchholz and Stephen Reich.

Case 1 – Headache

A 33 year old, right handed woman and current tobacco user comes to see you in the office for frequent headaches. The headaches started many years ago, but have been worse for the past 12 months. They occur twice a week on average. She can usually tell when a headache is coming because she feels nauseated and gets blurry, star shaped spots in her vision. Once the headache starts they last for many hours – typically she has to "sleep them off." Her sister has similar episodes. Otherwise, she is active, on oral medication for contraception and amlodipine for hypertension. She takes occasional over the counter ibuprofen for the headaches, with limited benefit.

1. What is the most appropriate diagnosis for this patient's headache type?
A. Retinal migraine
B. Tension headache
C. Migraine with aura
D. Abdominal migraine

2. Which of the following medication is most appropriate for aborting acute headaches of this type?
A. Acetaminophen
B. Butalbital + acetaminophen + caffeine (Fioricet)
C. Sumatriptan
D. Riboflavin + magnesium

3. Which of the following medications is the most useful adjunct for this patient's headache associated nausea?
A. Metoclopramide
B. Loperamide HCl (Imodium)
C. Bismuth subsalicylate
D. Esomeprazole magnesium

4. How should you council this patient in regards to how her headache history and habits affect her health risk?
A. Advise her that she has a higher than usual risk of endometriosis
B. Advise her that she has a higher than usual risk of miscarriage
C. Advise her that she has a higher than usual risk of stroke
D. Advise her that she has a higher than usual risk of gastrointestinal cancer

5. This patient has never had brain imaging. Is it indicated now?
A. Yes   B. No

Case 1 – Migraine Headache

This patient has typical migraine with visual aura. Migraine with aura leads to slightly higher risk for stroke compared to the general population (patients are also more likely to have a PFO). Smoking and using oral contraceptives significantly increase the risk of stroke associated with migraine headache with aura and these risk should be discussed with every migraine patient.

> **Typical Migraine Features:**
> - Duration of 4 – 72 hours
> - Two of the following four features:
>   - Unilateral, pulsatile, moderate to severe pain, aggravated by physical activity
> - Accompanied by nausea and/or photophobia
> - Auras, if present, are fully reversible visual, sensory or language symptoms with usually develop over several minutes and last no longer than 60 minutes

Brain imaging is usually reserved for patients with the following **red flags**:
- Headache beginning later in life (age 40 or greater)
- Thunderclap headache (onset to maximal pain within 1 minute)
- Exercise-induced or positional headache
- Significant change in headache frequency or severity
- HIV or cancer, other systemic illness
- Papilledema (swelling of the optic disc)
- Focal neurologic deficits, confusion, seizure
- Headache and signs of infection (fever, rash, stiff neck)

1. What is the most appropriate diagnosis for this patient's headache type?
C. Migraine headache with visual aura. Note that retinal migraines are a rare migraine type with monocular visual loss, and is a diagnosis of exclusion

2. Which of the following medication is most appropriate for aborting acute headaches of this type? **C. Sumatriptan – the triptans are effective when used at symptom onset**

3. Which of the following medications is the most useful adjunct for this patient's headache associated nausea? **A. Metoclopramide, can also use prochlorperazine or chlorpromazine**

4. How should you council this patient in regards to how her headache history and habits affect her health risk? **C. Advise her that she has a higher than usual risk of stroke**

5. This patient has never had brain imaging. Is it indicated now?
B. No, not if there are no red flags (see list above)

Acute Migraine Treatment

Simple Analgesics
- Acetaminophen and naproxen are appropriate for mild migraines, without nausea; can try IV ketorolac
- Best results when immediately taken at headache onset

Triptans
- Options include sumatriptan, rizatriptan, eletriptan, zolmitriptan, etc.
- Subcutaneous, oral, intranasal and transcutaneous options exist, but IM and intranasal are usually most efficacious
- Patients who don't respond to one triptan may respond to another
- Non-oral formulations best for patients with severe nausea
- Can cause vasospasm – contraindicated in patients with history of coronary artery disease or stroke

Antiemetics
- Metoclopramide 10 mg IV
- Prochlorperazine 10 mg IV or IM, or 25 mg PR
- Chlorpromazine 0.1 mg/kg IV – up to 25 mg IV
- Consider adding diphenhydramine (Benadryl) to the above antiemetics, as they all carry the risk of dystonic reaction

Other
- IV fluids important, dehydration common in migraine
- IV valproic acid 500 mg push (screen for pregnancy)
- IV magnesium 1 gram over 10-30 minutes, monitor heart rate (up to 5g total). Caution with renal impairment

Opioids and Barbiturates
- Should be avoided in acute migraine treatment
- Increased risk of developing chronic migraine and medication overuse headache
- Strongly advise against using butalbital (Fioricet)

For more information on migraine diagnosis, refer to the International Headache Classification website: www.ichd-3.org. Bookmark this site if you see headache patients frequently.

Chronic Migraine Treatment
Consider using preventive medications when patients have 1) three or more headaches per month, 2) headaches interfere with daily life, 3) acute medications are not effective.

Preventive medications usually require a trial of at least 2-3 months to determine efficacy. Remind patients that the goal is a 50% reduction in headache frequency. Ask patients to keep a headache diary to track frequency and severity of symptoms, which will help to identify headache trends over time. Medications may be used in combination if refractory, and medications may be tapered once migraine is well controlled. Topiramate is a good first choice medication.

Lifestyle Modification
- Many migraines respond to a healthier lifestyle, including increased exercise, better sleep and good hydration
- Eating protein in the mornings may help (improved serotonin availability in the brain)
- A sun lamp (light box) may help during the winter

Non-pharmacological
- Over the counter magnesium and riboflavin are helpful
- There is a wearable nerve stimulator called Cefaly, which is FDA approved for migraine treatment and prevention

Daily Preventive Medications
(there are many options – here are some with the best evidence)
- Propranolol 80 mg-240 mg or Metoprolol 50-150 mg daily
- Topiramate 25-150 mg daily (*a great choice, try 50 mg BID*)
- Nortriptyline or amitriptyline, low doses given QHS
- Divalproex sodium 250 mg-1500 mg daily
- Venlafaxine or duloxetine

Botox
- Patients must usually fail adequate trials of 2 or more oral medications
- Given subcutaneously every 3 months, botulinum toxin can be very effective for migraine prophylaxis

Migraine Aura

There are many different manifestations of migraine aura. Here are some of the most common:

Visual Aura is the most common aura. They often result in blurring of the vision. They may be black and white, zig-zag lines (Figure 1) or multicolored and pointy (Figure 2), the latter is often referred to a fortification scotoma.

▲ **Figure 1.** *Typical visual scotoma.*

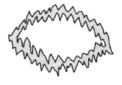
▲ **Figure 2.** *Fortification scotoma.*

Sensory Aura typically patients report a tingling sensation marching up the arm and into the face, developing over minutes. These is called "formication" because it's thought to feel like ants on the skin! Note that it is not a sensory loss, which would be more stroke like.

Language Aura many patients report subjective trouble word finding or getting their words out correctly. Objectively, this can be much less obvious.

Teaching Point: Auras lasting longer than 60 minutes are very suspicious, and deserve a brain MRI. Rarely, small infarcts can occur in the setting of prolonged migraine aura. However, if there are multiple sequential auras, each can last up to 60 minutes.

Aura symptoms are totally reversible. Typically aura symptoms are unilateral – i.e. the visual aura starts on one side of the visual field, or sensory symptoms slowly spread across one arm.

Evaluating the Headache Patient

Key History:
- Mode of onset (slow vs fast)
- Location
- Focal symptoms
- Exertion during onset or before
- Current medications and drug use
- Family history
- Medical illness
- Associated symptoms

Key Exam Steps:
- Vitals
- Eye exam including funduscopy
- Test for meningismus
- Head and neck exam
  - Temporomandibular region, sinuses, submandibular region, carotid arteries
- Complete neurologic exam

Do not assume that people with chronic migraines cannot have other headache types, including secondary headaches!

Medication Overuse Headaches

Patients who use analgesic medications on a frequent basis to treat headaches run the risk of developing a secondary headache type – medication overuse headaches, also known as rebound headaches. It is more likely to occur with opiates or medications such as butalbital, but is also common with NSAIDs and can even occur with triptans.

- Medication overuse can result in chronic, daily headaches – with headache often developing in the morning, and with headaches increasing in frequency over time as the analgesic is uptitrated.
- The patient must wean off the overused medication in order to break the cycle. This can be done be establishing a good preventive medication regimen and setting limits on the use of acute medications.
- Medication overuse headaches can occur with as little as 5 days per month of butalbital or 8 days per month of opiates, or 15 days of a combination of these with simple analgesics.
- The prognosis for medication overuse headache is good, if patients can abstain from overuse.

Case 2 – Headache

A 27 year old right handed woman with a history of obesity, asthma and polycystic ovarian disease presents to you complaining of several months of progressively worsening daily headaches. The pain is pulsatile, and present throughout the day. She complains of limited activity due to discomfort, and notes a recent 40 pound weight gain. She complains of episodic transient visual loss, especially when bending forward. On exam she is noted to have mild papilledema (swelling of the optic discs).

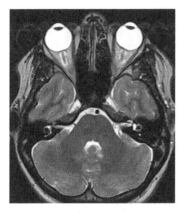

◄ *T2W brain MRI. The radiologist read it as showing "no acute intracranial pathology."*

1. The patient's symptoms are most concerning for which condition?
A. Depression and tension headache
B. Increased intracranial pressure
C. Progressive brain tumor
D. CNS vasculitis

2. What is the next best diagnostic step to confirm this condition, following the normal brain MRI?
A. Lumbar puncture with opening pressure
B. Trial of oral prednisone
C. Repeat brain MRI in 2 weeks
D. EEG

3. Which of the following medications is most useful for treating this patient's condition?
A. Verapamil
B. Phenobarbital
C. Procainamide
D. Acetazolamide

Case 2 – Idiopathic Intracranial Hypertension

This patient has idiopathic intracranial hypertension (IIH), formerly known as "pseudotumor cerebri." In this condition patients develop increased cerebrospinal fluid pressure, headaches and – in advanced cases – vision loss. Another possible neurological symptom is a 6th nerve palsy, causing diplopia.

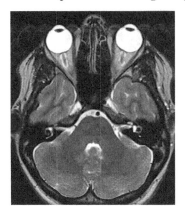

◄ *Brain imaging is normal in IIH, but is necessary to exclude a true intracranial mass or cerebral venous thrombosis.*

Patients are typically, but not always, obese. Symptoms may improve with weight loss, which is a first line treatment. The medication of choice is acetazolamide, which is both a diuretic and slows the formation of new cerebrospinal fluid. In extreme cases a procedure known as optic nerve sheath fenestration can help shunt CSF, and is usually done to preserve vision. Lumbar puncture may be helpful acutely to make the diagnosis and to relieve pressure if vision is being lost, but serial lumbar punctures are not usually done.

1. The patient's symptoms are most concerning for what condition?
B. Increased intracranial pressure, in this case it is idiopathic

2. What is the next best diagnostic step to confirm this condition, following the normal brain MRI?
A. Lumbar puncture with opening pressure – normal is < 20 mmHg

3. Which of the following medications is most useful for treating this patient's condition?
D. Acetazolamide, which may also decrease CSF production

> **Tip**: Certain medications are associated with the development of IIH, such as doxycycline and high doses of vitamin A (i.e. isoretinoin).

Case 3 – Headache

A 30 year old right handed man comes to you for multiple episodes of severe right sided facial pain, just above the right eye. Episodes last between 20 minutes to 2 hours each and occur between 3 and 8 times per day. He had a similar string of headaches one year ago, which resolved spontaneously after one month. Nothing helps, and he paces restlessly until the pain resolves. During the episodes he develops a red, watery right eye, a small right pupil and slight drooping of the right eyelid.

1. What is the most appropriate diagnosis for this patient's headache type?
A. Migraine with aura
B. Cluster headache
C. Trigeminal neuralgia
D. Paroxysmal hemicrania

2. What is the medication of choice to <u>prevent</u> these episodes?
A. Propranolol
B. Indomethacin
C. Verapamil
D. Carbamazepine

3. What is the only FDA approved treatment for <u>acute</u> episodes of this headache type?
A. DHE and sumatriptan
B. Piroxicam
C. Lamotrigine
D. Sublingual nitroglycerine

Case 3 – Cluster Headache

This patient has cluster headaches. Cluster headaches are one of many, uncommon but important, headache types known as the trigeminal autonomic cephalgias. These headache types often have accompanying parasympathetic signs, such as miosis, rhinorrhea, sweating or ptosis. They typically last less than 4 hours. The duration of the episode is a useful diagnostic feature. Imaging with contrast enhanced MRI is indicated for a new onset suspected trigeminal autonomic cephalgia.

1. What is the most appropriate diagnosis for this patient's headache type?
B. Cluster headache

2. What is the medication of choice to prevent these episodes?
C. Verapamil

3. What is the only FDA approved treatment for acute episodes of this headache type?
A. DHE and sumatriptan

Overview of the Trigeminal Autonomic Cephalgias

FEATURE	SUNCT	PAROXYSMAL HEMICRANIA	CLUSTER	HEMICRANIA CONTINUA
Male: Female	1.5:1	1:2	3:1	1:2
# Attacks/day	1-100	1-40	1-8	Daily in half of patients
Attack Duration	Seconds - 10 minutes	2-30 minutes	15-180 minutes	30 minutes – 3 days
Nocturnal?	No	No	Yes	No
Triggers	Neck movement	Cervical root pressure	EtOH, Nitro	None
Treatment	Lamotrigine	Indomethacin	O2, sumatriptan, verapamil	Indomethacin

Case 4 – Headache

A 27 year old left handed woman with frequent migraine headaches is 22 weeks pregnant. She has recently had an exacerbation of her baseline migraine with aura, and reports 3 - 4 migraines per week accompanied by photophobia and severe nausea. She is not taking any medication other than a prenatal vitamin for fear of safety.

1. All of the following medications are Category B in the first two trimesters, and are acceptable for limited use, <u>except for</u> which?
A. Acetaminophen
B. Sumatriptan
C. Dihydroergotamine (DHE)
D. Naproxen

2. Which of the following is the drug of first choice for prevention of tension type headaches in pregnancy:
A. Topiramate
B. Amitriptyline
C. Gabapentin
D. No preventive headache medications are recommended

3. The use of gadolinium contrast for MRI during pregnancy is _____.
A. safe and well tolerated
B. safe, but should be used judiciously
C. strictly contraindicated
D. known to cause midline defects, including cleft lip

4. A history of migraine headaches puts this patient at increased risk for which complication of pregnancy?
A. Eclampsia
B. Gestational diabetes
C. Premature labor
D. Premature rupture of membranes

Case 4 – Headache in Pregnancy

Headache treatment is limited in pregnant patients. Migraine frequency may improve or worsen, depending on the patient. Some women develop migraine with aura for the first time during pregnancy, regardless of their past history of headache. Non-pharmacological options such as stress management, increased exposure to morning sunlight or use of a sun lamp (i.e. SAD lamp) and light exercise can also be helpful adjuncts to headache control during pregnancy.

1. All of the following medications are Category B in the first two trimesters, except which? **C. Dihydroergotamine (DHE) – strongly contraindicated in pregnancy**

2. Which of the following is the drug of first choice for prevention of tension type headaches in pregnancy:
B. Amitriptyline – also consider verapamil or propranolol

3. The use of gadolinium contrast for MRI during pregnancy is _____.
C. strictly contraindicated

4. A history of migraine headaches puts this patient at increased risk for which complication of pregnancy?
A. Eclampsia and pre-eclampsia

Headache Medications during Pregnancy:

Preferred medications:
- Acetaminophen (Cat B) – may combine with codeine
- Naproxen/ibuprofen (Cat B during first two trimesters only)
- Promethazine or metoclopramide for nausea (Cat B)

✓ Although triptans can be used judiciously for migraines during pregnancy, DHE is category X and must be avoided
✓ History of migraines increases the risk of developing hypertensive disorders of pregnancy, such as eclampsia
✓ Gadolinium should be avoided in pregnancy but is not contraindicated during lactation

---

For further reading see: MacGregor EA. Headache in Pregnancy. *Continuum*. 2013; (20)1: 128-147.

Case 5 – Headache

A 35 year old right handed man with a history only notable for tobacco abuse has been well until about three weeks ago, when he developed new headaches. The headaches have become progressively worse, are constant and dull. He thinks they are worse early in the morning when he has been laying down. Lately, he has also begun to feel nauseated. His exam is notable only for mild bilateral papilledema (swelling of the optic discs). A contrast enhanced head CT is obtained which shows dural venous sinus thrombosis.

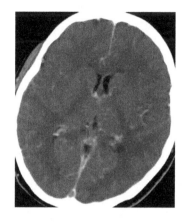

◄ *Contrast enhanced head CT.*

1. What is the treatment of choice for a patient with uncomplicated dural venous sinus thrombosis?
A. IV tPA
B. Aspirin + clopidogrel
C. Heparin infusion
D. Mechanical thrombectomy

2. Patients with venous thrombosis have a predisposition to brain hemorrhage. Which treatment is typically used when small amounts of cerebral hemorrhgae are present?
A. IV tPA
B. Aspirin + clopidogrel
C. Heparin infusion
D. Mechanical thrombectomy

3. If the patient complained of vision loss, which procedure could be performed prior to starting anticoagulation?
A. Lumbar puncture   C. Lumbar drain
B. Optic nerve sheath fenestration   D. Ventriculoperitoneal shunting

Case 5 – Dural Venous Sinus Thrombosis

This patient was found to have a dural venous sinus thrombosis, resulting in increased intracranial pressure. The superior sagittal sinus is the most commonly affected venous structure. Venous blood flow is impaired, causing back pressure and a tendency for a "low pressure" congestive venous hemorrhage. Even when mild hemorrhage is present, the treatment of choice is typically cautious heparin anticoagulation. Although practices differ, mechanical venous thrombectomy is also an option in select cases.

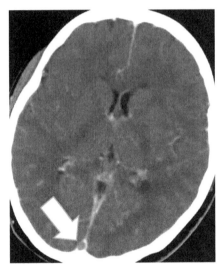

◀ *Contrast enhanced head CT showing a blood clot within the sagittal venous sinus (arrow). Contrast surrounds the clot, creating the "empty delta" sign (below).*

1. What is the treatment of choice for a patient with uncomplicated dural venous sinus thrombosis?
C. Heparin infusion

2. Patients with venous thrombosis have a predisposition to brain hemorrhage, in which case which treatment is typically used?
C. Heparin infusion, often without bolus and conservative PTT goal

3. If the patient complained of vision loss, which procedure could be performed prior to starting anticoagulation?
A. Lumbar puncture – relieving ICP may help preserve vision

Tip: Prothrombotic conditions are often present, including pregnancy, dehydration, occult neoplasm, oral contraceptive or tobacco use, etc.

Case 6 – Headache

A 55 year old right handed woman with long standing hypertension experienced severe, progressive occipital headache over the past two days. On the day of presentation she complained to her spouse about vision loss and then subsequently developed confusion, followed by a generalized tonic-clonic seizure. On presentation her blood pressure was 219/113 mmHg and her exam was non-focal, but she was confused and vision remained impaired. Her non-enhanced brain MRI from the emergency department is shown here.

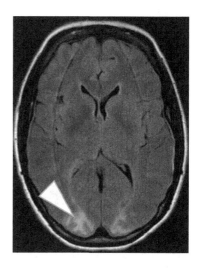

1. What is the abnormality identified by the arrowhead on the brain MRI?
A. Infection
B. Stroke
C. Cerebral edema
D. Brain tumor

2. The mainstay of treatment for this condition involves (choose the best answer):
A. Supportive care in the ICU
B. Treating the acutely elevated blood pressure
C. Starting IV antiviral medications
D. Surgical resection and radiotherapy

3. Development of this condition is associated with use of all of the following medications except which?
A. Tacrolimus (Prograf)
B. Bevacizumab (Avastin)
C. Mycophenolate mofetil (Cellcept)
D. Cyclosporine (Neoral)

Case 6 – Posterior Reversible Encephalopathy Syndrome (PRES)

This patient has Posterior Reversible Encephalopathy Syndrome, known as PRES. It is a form of hypertensive emergency, in which patients develop cerebral edema from elevated blood pressures. The primary treatment is lowering the blood pressure, although patients may develop seizures which also require treatment. The symptoms of PRES typically include severe headache, and may go on to cause stroke or brain hemorrhage. It is not always reversible. In addition, it is not always limited to the occipital cortex.

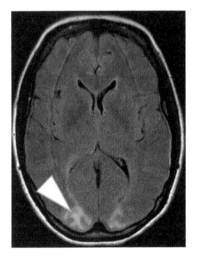

◄ *T2W FLAIR brain MRI, with typical bilateral FLAIR hyperintensities (arrowhead) consistent with PRES. Notice how the signal is confined to the white matter, and spares the cortex – this is typical of vasogenic edema.*

Patients on immune modulating medications, especially cyclosporine and tacrolimus, are prone to developing PRES at even lower blood pressures.

In addition, the cerebral edema from PRES is not always located in the bilateral occipital lobes, but can be in the brainstem, or bilateral frontal lobes.

1. What is the abnormality identified by the arrowhead on the brain MRI? **C. Cerebral edema**

2. The mainstay of treatment for this condition involves (choose the best answer): **B. Treating the acutely elevated blood pressure**

3. Development of this condition is associated with use of all of the following medications except which? **C. Mycophenolate mofetil – tacrolimus, bevacizumab and cyclosporine all raise the risk of developing this condition, sometimes at lower blood pressures**

Case 7 – Headache

A 28 year old man of unknown handedness is brought to the emergency department by his friends after three days of progressive headache, bizarre behavior and forgetfulness. This morning, he developed a fever of 39° C. Just before arriving in the hospital the patient had a generalized tonic-clonic seizure. On exam he is lethargic, confused and sweaty. His left arm appears weak and he strongly resists any movement of his neck.

1. This patient needs a lumbar puncture (LP). Should a head CT be performed prior to this procedure?
A. Yes, a head CT prior to LP is required for all suspected meningitis
B. Yes, for patients with focal deficits including seizure
C. No, the lumbar puncture should not be delayed
D. No, a brain MRI must be done prior to the LP

2. For patients with suspected bacterial meningitis, which medication should be administered prior to, or at the same time as, starting antibiotics?
A. Hypertonic saline
B. Epinephrine
C. Mannitol
D. Dexamethasone

3. The lumbar puncture reveals a lymphocytic predominant pleocytosis with 250 cells/mm³ and CSF protein of 80 mg/dl. CSF glucose is low normal. The non-enhanced head CT shown here is obtained. Which organism do you now most strongly suspect?
A. *Meningococcus*
B. *H. influenza*
C. HSV-1
D. HSV-2

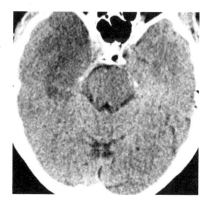

Case 7 – Herpes Simplex Virus Meningoencephalitis

The patient presented with fever, headache and altered mental status – the classic triad for meningoencephalitis. Not every patient has all 3 symptoms; you must have a low threshold of suspicion. In this case, the lymphocytic pleiocytosis and temporal lobe involvement should strongly suggest that this is a case of herpes encephalitis. In adults, the cause is almost always HSV-1. Milder cases can be caused by HSV-2. The treatment of choice for herpes encephalitis is IV acyclovir, administered with IV fluid for renal protection.

Dexamethasone 0.15 mg/kg q6 hours is given when bacterial meningitis is suspected – and only continued if *S. pneumonia* is the infective organism.

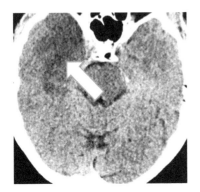

◄ *Non-enhanced head CT showing right temporal lobe hypodensity and expansion (arrow) – very suggestive of herpes encephalitis in this case.*

1. This patient needs a lumbar puncture (LP). Should a head CT be performed prior to this procedure? **B. Yes, for patients with focal deficits including seizure. Intracranial masses must be ruled out before LP due to the risk of herniation.**

2. For patients with suspected bacterial meningitis, which medication should be administered prior to, or at the same time as, starting antibiotics?
D. Dexamethasone – shown to decrease morbidity for *S. pneumonia*

3. The lumbar puncture reveals a lymphocytic predominant pleiocytosis with 250 cells/mm^3 and CSF protein of 80 mg/dl. The following non-enhanced head CT is obtained. Which organism do you now most strongly suspect?
C. HSV-1, which classically involves the frontal and temporal lobes

When to obtain a head CT prior to LP in meningitis:

✓ When focal neurologic symptoms are present
✓ New onset of seizure
✓ Immunocompromised state (i.e. HIV)
✓ There is a history of CNS mass lesion (i.e. cancer)
✓ Abnormal level of consciousness

Notes

Paul D. Johnson, MD

Sleep Disorders

"Would you like a nightcap?"
- "No, thank you. I don't wear them."
The Naked Gun: From the Files of Police Squad!

Sleep Disorders - Sleep Basics

There are four stages of sleep, stages 1-3 and REM (Rapid Eye Movement) sleep. Stage 3 is considered slow-wave sleep. In a typical night there are 4-6 cycles of non-REM sleep, each followed by a cycle of REM sleep. The initial cycle of REM starts about 90 minutes after sleep onset.

During normal REM sleep motor activity is suppressed – in effect, you are paralyzed. This keeps you from acting out your dreams, which also occur during REM sleep! The stage of sleep can be identified by EEG, as there are characteristic brain wave patterns for each of the four stages of sleep.

Sleep disorders are common. They affect 25 – 30% of all adults.

Components of a good sleep history:

- ✓ Sleep related breathing disorders
    - Snoring, morning headache, witnessed apneas
- ✓ Nocturnal behaviors
    - Bruxism, leg movement, enacting dreams, sleep walking
- ✓ Nocturnal awakenings
    - Including leg discomfort, urge to move
- ✓ Sleep hygiene
    - Bed time, wake and rise (out of bed) times, bedtime routine including screen time, pets in bed, etc.
- ✓ Daytime functioning
    - Mood, impaired performance, concentration & memory
- ✓ Use of sleep aids & stimulants
    - Caffeine use, sedative use

Ways to Evaluate Sleep

Epworth Sleepiness Scale (ESS) - 24 point questionnaire to identify patients with elevated levels of daytime sleepiness.

Polysomnogram (PSG) - multimodal sleep study used to evaluate many types of sleep disorders – often requires an overnight stay in a sleep lab.

Multiple Sleep Latency Test (MSLT) - sleep test that evaluates sleep latency, or time it takes to fall asleep. Used primarily to evaluate for narcolepsy.

Sleep Disorders - Parasomnias

Abnormal behaviors that occur during sleep, known as parasomnia, are differentiated by which stage of sleep they occur during - either REM sleep, non-REM sleep or during sleep-wake transitions.

Sleep Disorder	Sleep Stage	High Yield Facts
Nightmares	REM	Occurs during last 1/3 of night, dream is remembered in detail
Cluster headache	REM	Severe headache with autonomic signs - treat with high flow oxygen
Rhythmic movement d/o	Sleep/wake transitions	Often head-banging or similar behavior, occurs in children
PLMD*	Stage 1 & 2	Involuntary, stereotyped leg flexion
Sleep walking	Stage 3	More common in first 1/3 of sleep, patient amnestic to complex behaviors - may be familial
Night terrors	Stage 3	Occur early in the night, no recollection of events by child, autonomic symptoms
Confused arousals	Stage 3	Child awakens confused and disoriented

▲ *Parasomnias and when they occur. Note that all events occurring in stages 1-3 are non-REM events. * PLMD = periodic leg movement disorder.*

Other common nocturnal events that are not parasomnias:

Nocturnal enuresis (i.e. bed wetting), occurs more frequently in boys and is normal until about age 5. Other than behavioral treatments, such as scheduled urination and limiting fluid intake, imipramine and DDAVP can be used as treatment.

Another nocturnal event that causes consternation is **benign neonatal myoclonus**, which occurs in otherwise healthy infants. The myoclonus disappears upon awakening and is not related to epilepsy.

Bruxism or teeth-grinding is also a common, non-REM phenomenon, affecting about 80% of people with a peak prevalence in young adults.

Case 1 – Sleep Disorders

A 65 year old right handed man presents with complaints of leg jerking that keeps his wife awake and occasional awakens him from sleep. He endorses significant daytime fatigue, and has even fallen asleep in the car while waiting at a stop light. His wife notes that he snores very loudly and has seen him stop breathing altogether during the night – although she does continue to see his chest rise during these events. The patient is obese and has refractory hypertension. While he denies tobacco use he does have two beers each night before bed, and drinks large amounts of coffee during the day.

1. Which assessment could be performed as a measure of this patient's daytime sleepiness?
A. Montreal Cognitive Assessment (MoCA)
B. Barthel Index
C. FIM scale
D. Epworth sleepiness scale (ESS)

2. You suspect that this patient most likely has which sleep related disorder?
A. Central sleep apnea
B. Obstructive sleep apnea
C. Complex sleep apnea syndrome
D. Central hypoventilatory syndrome

3. Diagnosis of obstructive sleep apnea is based primarily on which test?
A. Multiple sleep latency test (MSLT)
B. EEG
C. Polysomnogram (PSG)
D. Maintenance of wakefulness test (MWT)

4. How long must a patient's pause in breathing last in order to be classified as an apneic event?
A. > 3 seconds
B. > 5 seconds
C. > 10 seconds
D. > 15 seconds

5. Which of the following is <u>not</u> a risk factor for apneic events?
A. Obesity
B. Evening alcohol use
C. Male sex
D. Refractory hypertension

Case 1 – Obstructive Sleep Apnea

This patient has **obstructive sleep apnea** (OSA). Risk factors include male sex, obesity – especially larger neck circumference and older age. Sedative use before bed makes apneic episodes more likely. OSA is due to collapse of the upper airway, preventing ventilation during sleep. It results in fragmented sleep, frequent brief awakenings called "micro-arousals" and increased blood pressure.

OSA is diagnosed with an overnight sleep study, called a polysomnogram, and based on the number of apneas (cessation in breathing lasting 10 seconds) or hypopneas (shallow breathing) which occur. This is calculated as the Apnea/Hypopnea Index or AHI, and is a measure of severity of OSA. The normal AHI is ≤ 5 per hour.

There are several treatment options for OSA. Continuous positive airway pressure (CPAP) is the most common, but surgical options are also available. Patients often see improvement by sleeping on their side or sleeping sitting up.

Central sleep apnea is a condition where night time sleeping is impaired due to decreased respiratory drive from the brainstem. It is more common after a stroke or brain injury, and is treated differently than OSA.

The ESS measures the severity of daytime sleepiness. The FIM and Barthel index are measurements of disability, usually performed by rehab specialists.

1. Which assessment could be performed as a measure of this patient's daytime sleepiness? **D. Epworth sleepiness scale – the FIM and Barthel index are measurements of disability, often used after stroke**

2. You suspect that this patient most likely has which sleep related disorder?
B. Obstructive sleep apnea

3. Diagnosis of obstructive sleep apnea is based primarily on which test?
C. Polysomnogram (PSG)

4. How long must a patient's pause in breathing last in order to be classified as an apneic event?
C. > 10 seconds

5. Which of the following is not a risk factor for apneic events?
D. Refractory hypertension – although OSA might cause refractory hypertension

Case 2 – Sleep Disorders

A 17 year old boy is evaluated for excessive daytime sleepiness. He has difficulty waking up each morning and is frequently late to school. At night he is unable to fall asleep, despite getting into bed at a reasonable time. Over the weekend he stays up until late at night and sleeps late into the day, awakening feeling fully refreshed. He denies using any electronics prior to sleep.

1. What is the most likely diagnosis for this patient's difficulty waking up in the mornings on week days?
A. Primary insomnia
B. Advanced sleep phase syndrome
C. Delayed sleep phase syndrome
D. Circadian dysrhythmia

2. What changes typically occur in sleep architecture in this condition?
A. Decreased REM duration
B. Decreased sleep latency
C. Fragmented REM sleep
D. None – there is normal sleep architecture

Case 2 – Circadian Rhythm Disorders

This teenager has delayed sleep phase syndrome, a common and transient sleep disorder often seen in adolescents. The patient's circadian rhythm does not fit the schedule imposed by society, and you note that he feels refreshed after sleeping in on weekends. Patients who awaken early and fall asleep early have an advanced sleep phase syndrome.

Circadian rhythm disorders are very common, and include jet lag and shift worker syndrome.

Stress and anxiety related to not getting enough sleep can lead to a psychophysiologic insomnia, in which patients spend so much time worrying about their lack of sleep that it keeps them awake at night.

1. What is the most likely diagnosis for this patient's difficulty waking up in the mornings on week days?
C. Delayed sleep phase syndrome

2. What changes typically occur in sleep architecture in this condition?
D. None – there is normal sleep architecture

> **Commonly used medications for insomnia:**
>
> - **Melatonin** great first line sleep aid, least amount of side effects
> - **Zolpidem** can cause amnesia and sleep walking
> - **Clonazepam** a potentially habit forming benzodiazepine, can leave patients unsteady & with cognitive impairment
> - **Trazodone** rarely causes priapism, myoclonus
> - **Quetiapine** an antipsychotic that really should be <u>avoided</u> for sleep alone!
> - **Diphenhydramine** found in all sorts of over the counter sleep aids, can cause falls, cognitive impairment and urinary retention

Teaching Point: A good first line sleep aid, especially in the elderly, is over the counter melatonin, which is a well-tolerated supplement – although not FDA approved. A recommended starting dose is 3 mg taken a few hours prior to sleep.

Case 3 – Sleep Disorders

A 25 year old right handed man reports frequently dropping things whenever he is excited or after he hears a good joke. In addition, he finds himself drinking 6 or more cups of coffee daily just to stay awake. In general he sleeps well, getting a full 8 hours of sleep. Despite that he frequently falls asleep at work. He denies snoring and has no witnessed apnea, but he does note several occasions where he was temporarily unable to move upon awakening.

1. Diagnosis of this patient's condition could best be made by which test?
A. Multiple sleep latency test (MSLT)
B. Polysomnogram (PSG)
C. Cerebrospinal fluid analysis (CSF analysis)
D. Maintenance of wakefulness test (MWT)

2. Which of the following is the most specific symptom for this patient's condition?
A. Excessive daytime sleepiness & sleep attacks
B. Hypnogogic hallucinations
C. Sleep paralysis
D. Transient episodes of weakness, provoked by emotion

3. Which of the following is the most well-established treatment for the patient's complaint of frequent weakness?
A. Dextroamphetamine
B. Pramipexole
C. Pregabalin
D. Gamma hydroxybutyrate (GHB)

4. What of the following findings on a multiple sleep latency test is most consistent with a diagnosis of narcolepsy?
A. Inability to awaken the patient during the test
B. Sleep latency < 5 minutes
C. Sleep latency < 10 minutes
D. Absence of REM sleep

5. Does this patient require brain imaging?
A. Yes – head CT   B. Yes – brain MRI   C. No

Case 3 – Narcolepsy with Cataplexy

The patient presents with narcolepsy with cataplexy. Narcolepsy is a REM sleep disorder of excessive sleepiness despite getting adequate sleep. The typical patient will fall asleep within 5 minutes of undisturbed rest. Narcolepsy can occur with or without cataplexy. The features of cataplexy are weakness, which can manifest as subtle head nodding, arm weakness, slurred speech of clumsiness, usually triggered by strong emotion such as laughter or fear.

Patients with narcolepsy often have subtle hallucinations upon falling asleep (called hypnagogic hallucinations), or brief periods of paralysis upon awakening – however, both of these features are also seen in otherwise healthy people.

Cataplexy responds well to gamma hydroxybutyrate (GHB), but can also be treated with the antidepressant SSRI medications or with imipramine. Stimulants such as dextroamphetamine or modafinil are useful for the excessive sleepiness of narcolepsy.

1. Diagnosis of this patient's condition could best be made by which test?
A. Multiple sleep latency test (MSLT)

2. Which of the following is the <u>most specific</u> symptom for this patient's condition?
D. Transient weakness provoked by emotion, known as cataplexy

3. Which of the following is the most well-established treatment for the patient's complaint of frequent weakness?
D. Gamma hydroxybutyrate (GHB), although this can be difficult to obtain

4. What of the following findings on a multiple sleep latency test is most consistent with a diagnosis of narcolepsy?
B. Sleep latency < 5 minutes

5. Does this patient require brain imaging?
C. No – not if the patient has a classic history of narcolepsy with cataplexy

Case 4 – Sleep Disorders

A 53 year old right handed woman comes to clinic with a complaint of poor sleep. She feels that falling asleep is very difficult due to the development of a strong urge to move her legs which occurs almost every evening. When she tries to lay still, she develops an "tingly, creepy crawly" feeling in both legs which is only relieved by getting up and walking – which completely relieves the sensation. She never experiences these episodes during the day. She denies joint pain, and massaging the legs does not help.

1. Which one lab test is most important when evaluating this condition?
A. Thyroid stimulating hormone levels
B. Hemoglobin A1c
C. Serum ferritin
D. Vitamin B12 levels

2. This condition is commonly seen in patients with which other sleep disorder?
A. Obstructive sleep apnea
B. Narcolepsy
C. Sleep terrors
D. Central sleep apnea

3. All of the following are effective treatments for this condition except which?
A. Pramipexole
B. Caffeine
C. Carbidopa/levodopa
D. Gabapentin

4. Which of the following is the most likely diagnosis of this patient's condition?
A. Restless leg syndrome
B. Diabetic polyneuropathy
C. Tardive dyskinesia
D. Anxiety

Case 4 – Restless Leg Syndrome

This patient has classic symptoms of restless leg syndrome (RLS). It is a clinical diagnosis, and no confirmatory testing is required – however, low serum iron levels can make the condition worse. Serum ferritin should be checked in patients with chronic RLS, and iron supplementation combined with vitamin C should be administered if iron levels are in the low normal range or below.

There are many medical treatments available for RLS, including carbidopa/levodopa, gabapentin and ropinirole or pramipexole. Keep in mind that an uncommon side effect of ropinirole and pramipexole is impulse control disorder, which can include gambling, compulsive shopping or impulsive sexual acts. RLS also has a tendency to worsen over time and require escalating doses of medical therapy.

Patients that you suspect of having RLS should also be screened for leg cramps, venous stasis, painful peripheral neuropathy and arthritis.

1. Which one lab test is most important when evaluating this condition?
C. Serum ferritin – low serum ferritin worsens RLS

2. This condition is commonly seen in patients with which other sleep disorder?
A. Obstructive sleep apnea, and may improve if OSA is treated

3. All of the following are effective treatments for this condition <u>except</u> which?
B. Caffeine – Pramipexole, gabapentin and carbidopa/levodopa are all effective treatments for RLS

4. Which of the following is the most likely diagnosis of this patient's condition?
A. Restless leg syndrome

Sleep Hygiene

- Bed time and rise time, enough for 7-8 hours of sleep?
- Sleep latency – how long does it take to fall asleep & why?
- Limit fluids, caffeine, exercise, TV/screen time in the evening
- Minimize nocturnal disruptions: bed partners, pets in the bed, cell phone alarms, text messages
- Reasons for nocturnal awakenings? Bathroom, pain, etc.

Notes

Paul D. Johnson, MD

Movement Disorders

"Sixty percent of the time, it works every time."
Anchorman: The Legend of Ron Burgundy

Neurologic Control of Movement

Movement is an extremely complex task, and it takes a lot of brain power to plan and execute fine motor movements, maintain balance and make movement seem effortless. Several key brain systems are involved. Although listed separately here, they all "talk" to each other and work closely together.

- **Corticospinal Tract** This is the pathway from the brain's motor cortex, running through the internal capsule, spinal cord and down to the motor neurons in the spinal cord. Injury in the CST leads to frank weakness, as we saw in the stroke chapter.
- **Basal Ganglia** These deep nuclei have a large role in facilitating involuntary movements and postural activities – diseases affecting the basal ganglia lead to problems regulating movements, either slowing and freezing as in Parkinson disease, or excessive and erratic movements, called dyskinesias.
- **Cerebellum** Responsible for coordinating and "smoothing" out movements, injury to the cerebellum leads to tremor, over shooting or under shooting (dysmetria) with intentional movements.

A major focus of this section will be diseases of the basal ganglia, the best example of which is Parkinson disease – which is due to degeneration of one key part of the basal ganglia, the substantia nigra. The substantia nigra produces a special neurotransmitter known as dopamine, which facilitates the working of the rest of the basal ganglia. If you understand Parkinson disease, you will be in a good position to understand other movement disorders resulting from diseases of the basal ganglia.

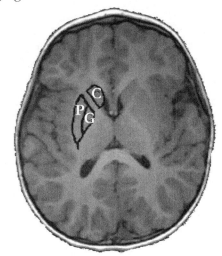

▶ *T1W MRI. Key parts of the basal ganglia are outlined here. G) Globus Pallidus, P) Putamen and C) the Caudate nucleus. There are several other smaller nuclei, not seen on this view. The most important of which is the substantia nigra, which is located in the midbrain.*

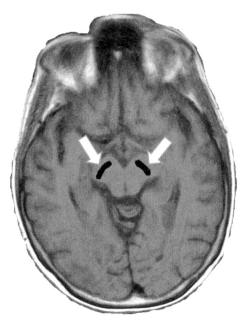

◢ *T1W MRI. The approximate location of the substantia nigra in the midbrain is shown here (arrows). This is the dopamine producing part of the brain, the loss of which causes problems throughout the basal ganglia system. This view is just slightly below the MRI on the previous page.*

Neurotransmitters:

- **D1 receptors** dopamine receptors that <u>increase</u> motor activity
- **D2 receptors** dopamine receptors that <u>decrease</u> motor activity
- **Glutamate** major <u>excitatory</u> neurotransmitter of the basal ganglia
- **GABA** is the major <u>inhibitory</u> neurotransmitter

Toxins

- **MPTP** a designer drug, causes severe parkinsonism
- **Carbon Monoxide** causes necrosis of the Globus Pallidus
- **Methanol** causes necrosis of the putamen

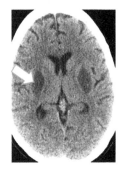

◂ *Non-enhanced head CT from a patient with methanol intoxication, showing selective necrosis of the bilateral putamen (arrow).*

Case 1 – Movement Disorders

A 60 year old left handed man is evaluated for progressive left sided tremors developing over the past year. The tremor is present at rest, and improves when reaching for or holding objects. His wife thinks he is becoming depressed, because he rarely smiles anymore. Other than constipation, a recent fall and dry eyes he has no major medical issues. On exam his left arm is stiffer than the right with passive movements. He walks slowly, with short steps, but does not fall in the clinic hallway.

1. All of the following are risk factors for developing Parkinson disease except for which?
A. Chronic exposure to well water (heavy metals)
B. History of significant head trauma
C. Tobacco use
D. Exposure to pesticides (ex. Agent Orange)

2. Should patients with suspected Parkinson disease undergo brain MRI?
A. Yes   B. No   C. Only if they have atypical features

3. All of the following are red flags in patients with Parkinson disease except for which?
A. Positive Babinski sign
B. Early development of orthostatic hypotension
C. Complete paralysis of vertical gaze
D. Loss of sense of smell

4. Which of the following medications can cause symptoms mimicking Parkinson disease?
A. Metoclopramide
B. Thalidomide
C. Linezolid
D. Cefepime

5. Which of the following is not a typical feature of Parkinson disease?
A. Tremor
B. Rigidity
C. Seizure
D. Akinesia (slow movements)
E. Postural instability

Case 1 – Early Parkinson Disease

This patient is showing signs of early Parkinson disease (PD). PD is typically unilateral at onset and gradually becomes bilateral. Although most PD is sporadic (also called idiopathic), a small number of cases are genetic. The incidence of PD increases sharply after age 60. Most, but not all, patients present with a prominent resting tremor. The most important features for diagnosis are the rigidity and slow movements (akinesia or bradykinesia).

Parkinson Disease Features
- Tremor – worse at rest
- Rigidity – stiffness in the limbs with passive movements
- Akinesia – slow movements
- Postural instability – falls and stooped gait are common

Other common (and early) findings in PD include loss of sense of smell, constipation, or dry eyes from limited blinking. Patients may get skin rashes such as seborrheic dermatitis.

Typically, anyone with suspected PD should have a brain MRI to exclude other pathology, such as vascular malformation or tumor near the basal ganglia.

Because PD is due to a loss of dopamine producing cells, any medications which block the action of dopamine can produce PD like symptoms – called "parkinsonism." Common medications causing parkinsonism include antiemetics such as metoclopramide, as well as antipsychotic medications like haloperidol or olanzapine.

1. All of the following are risk factors for developing Parkinson disease except for which?
C. Tobacco use – tobacco and caffeine both have a protective effect

2. Should patients with suspected Parkinson disease undergo brain MRI?
A. Yes, to rule out other causes of parkinsonism

3. All of the following are red flags in patients with Parkinson disease except which?
D. Loss of sense of smell – this is a very common early feature of PD

4. Which of the following medications can cause symptoms mimicking Parkinson disease?
A. Metoclopramide (a dopamine antagonist)

5. Which of the following is not a typical feature of Parkinson disease?
C. Seizure – there is no increased risk of seizures in PD

Case 2 – Movement Disorders

A 50 year old left handed man is evaluated for slowly progressive tremors involving both arms. Symptoms began over a decade ago. The tremors are present at rest, but become more apparent with either activity or stress. It is becoming difficult for him to eat or drink without spilling on himself, and he now struggles to sign his name legibly. Otherwise he is active, denies falls or changes in gait, and has no other medical issues. He had an uncle with similar tremors. A Dopamine Transporter (DaT) scan was obtained.

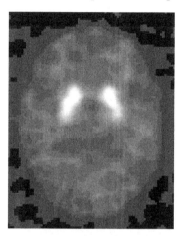

◄ *A DaT scan, which uses a small amount of a radioactive substance, called Ioflupane 123, to detect brain dopamine. This DaT scan shows the expected 'comma shaped' appearance of the dopamine producing part of the brain, the substantia nigra, and was read as normal.*

1. What is the most likely cause of this patient's tremor?
A. Parkinson disease
B. Essential tremor
C. Enhanced physiologic tremor
D. Fahr syndrome

2. Which of the following would be the <u>best</u> medicine to treat this patient's tremor?
A. Pimozide
B. Primidone
C. Carbidopa-Levodopa
D. Ropinirole

3. All of the following medications are associated with secondary tremor, except for which?
A. Amiodarone
B. Tacrolimus
C. Rituximab
D. Bupropion

Case 2 – Essential Tremor

This patient has an essential tremor – a benign condition although the tremor can be disabling. The major differential diagnosis for essential tremor is early Parkinson disease. A normal DaT scan essentially excludes a diagnosis of Parkinson disease. However, most clinicians will choose to follow a patient with serial examinations to monitor for progression of Parkinson disease symptoms, rather than order an expensive DaT scan. In this case, the decade long history of slow progression, the symmetry and the family history all support a diagnosis of essential tremor.

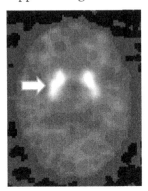

◀ *Normal DaT scan, showing the presence of dopamine in substantia nigra (arrow). In Parkinson disease these would be absent or much smaller. A DaT scan can help differentiate Parkinson disease from essential tremor or other acquired tremors.*

Essential Tremor	Parkinson Disease
▪ Slowly progressive	▪ Slowly progressive
▪ Action predominant	▪ Rest predominant
▪ Usually bilateral and symmetric	▪ Begins unilaterally
▪ Often familial	▪ Rarely familial
▪ No rigidity	▪ Rigidity prominent
▪ Head & neck involved in advanced cases	▪ Axial involvement is rare
Treatment for Essential Tremor	
▪ Primidone	▪ Deep Brain Stimulation (DBS)
▪ Propranolol	
▪ Avoiding stimulants, such as caffeine	
▪ Using weighted utensils	

1. What is the most likely cause of this patient's tremor?
B. Essential tremor – given the bilateral, slowly progressive nature

2. Which of the following would be the <u>best</u> medicine to treat this patient's tremor?
B. Primidone – a sedating medicine, it is a barbiturate

3. All of the following medications are associated with secondary tremor, except?
C. Rituximab – tacrolimus, bupropion, lithium & amiodarone cause tremor

Case 3 – Movement Disorders

An elderly woman with long standing Parkinson disease notes that she has gained almost 20 pounds, and racked up a credit card debt of almost $15,000 in the last three months. She never had a problem with gambling or binge eating until recently. When asked, she also complains of episodes of sudden fatigue. Three months ago she started a new medication for her Parkinson disease.

1. Which of the following medications was this woman most likely recently started on?
A. Dopamine agonist (ropinirole)
B. COMT inhibitor (entacapone)
C. Apomorphine
D. MAO-B inhibitor (selegiline)

2. Which of the following is another potential side effect of the medication that the patient recently started?
A. Myoclonic jerks
B. Increased hallucinations
C. Nausea
D. Central sleep apnea

3. Common side effects of carbidopa/levodopa (sinemet) include all of the following except:
A. Nausea
B. Dyskinesia
C. Hypotension
D. Rash

4. When taking carbidopa/levodopa, patients should be warned not to take the medication _____.
A. with food containing protein
B. with food containing carbohydrates
C. with grapefruit juice
D. with NSAIDs

Case 3 – Parkinson Disease Medications & Side Effects

This patient developed an impulse control disorder, a potential side effect of the **dopamine agonist** medications ropinirole and pramipexole.

Parkinson Disease Medications

Carbidopa/Levodopa (sinemet)
- Start at 50-100 mg/day divided into 3-4 doses
- Absorption impaired by protein – space out from meals
- Can cause nausea, hypotension, hallucinations, sexually inappropriate behavior, and over time leads to dyskinesias

COMT Inhibitors (entacapone)
- Prolong the half-life of carbidopa/levodopa
- Tablet may be mixed with sinemet (Stalevo), but has increased risk of nausea and dyskinesia

MAO-B Inhibitors (rasagiline/selegiline)
- Prolong the half-life of carbidopa/levodopa
- May delay the need to start levodopa

Dopamine Agonists (ropinirole/pramipexole/rotigotine)
- Widely used as adjunctive or monotherapy
- Rotigotine is a once-daily patch
- Risk of impulse control disorders; hyper sexuality, gambling, shopping, and over-eating.

Amantadine
- Anticholinergic medication that helps reduce dyskinesias

Sudden discontinuation of dopaminergic medications can precipitate **neuroleptic malignant syndrome**. This life threatening condition causes confusion, fever, rigidity and autonomic dysfunction. It is more often seen with antipsychotic (i.e. *anti*-dopaminergic) medications, but abrupt discontinuation of dopamine containing medication can have the same result.

1. Which of the following medications was this woman most likely recently started on? **A. Dopamine agonist (ex. ropinirole or pramipexole). COMT and MAO-B inhibitors both delay the breakdown of dopamine in the body, prolonging the effect of carbidopa/levodopa**

2. Which of the following is another potential side effect of the medication that the patient recently started? **B. Increased hallucinations – hallucinations become common late in PD. Patients often report seeing people or animals, especially in dim light.**

3. Common side effects of carbidopa/ include all of the following except: **D. Rash**

4. When taking carbidopa/levodopa, patients should be warned not to take the medication: _____ **A. with food containing protein, as it impairs absorption of the medication**

Case 4 – Movement Disorders

A 68 year old right handed man has had Parkinson disease for almost a decade. While previously well controlled on carbidopa/levodopa, he has recently had increasing problems with severe "on/off" phenomenon - he swings rapidly between having good control of his movements to being nearly frozen between doses of medication. In addition, when his carbidopa/levodopa dose is increased to overcome the freezing episodes, he develops uncontrolled writhing movements in his hands and neck. Although these don't bother him, they are very disturbing to his wife. The patient decided to undergo implantation of a deep brain stimulator to treat his advanced Parkinson disease symptoms – after which he noted significant improvement in his "on/off" periods.

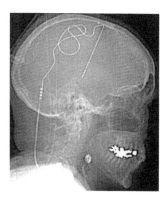

◀ *Radiograph showing an implanted deep brain stimulator (DBS) and leads.*

1. In advanced Parkinson disease all of the following are common <u>except</u> which?
A. Uncontrollable episodes of laughter and crying
B. Gait freezing
C. Medication related dyskinesias (uncontrolled abnormal movements)
D. Memory loss and dementia

2. Patients with a deep brain stimulator are unable to undergo brain MRI:
A. True   B. False

3. A deep brain stimulator causes an irreversible brain injury:
A. True   B. False

4. The major risk of placing a deep brain stimulator is:
A. Infection   C. Causing intractable epilepsy
B. Worsening of PD   D. Precipitating dementia

Case 4 – Advanced Parkinson Disease

Parkinson disease is a progressive neurodegenerative condition. As a result, patients have symptom progression over time. Movements become difficult to control with medication, with short treatment times followed by episodes of freezing. As doses of carbidopa/levodopa are increased patients may experience dyskinesias, an almost paradoxical effect of uncontrolled movements from excess dopamine. Lastly, many patients will develop cognitive issues and visual hallucinations as PD progresses.

Deep Brain Stimulation (DBS) can be a very effective treatment for select patients with moderate PD whose disease is not well controlled with medication. It involves surgical placement of electrodes into the subthalamic nucleus (part of the basal ganglia). The DBS can be programed and adjusted in clinic to provide constant symptomatic relief. This allows a lower dose of medication to be used, mitigating side effects. Patients must be otherwise healthy, with good cognitive function and no psychiatric issues to qualify for DBS placement.

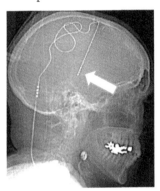

◄ *Radiograph showing an implanted deep brain stimulator (DBS). The electrode (arrow) is placed in the subthalamic nucleus and the battery is placed under the skin of the chest. The settings of the DBS are fully adjustable.*

Alternatively, amantadine 100 mg daily for 1 week then 100 mg BID can be used to try to reduce carbidopa/levodopa related dyskinesias.

1. In advanced Parkinson disease all of the following are common <u>except</u> which?
A. Uncontrollable episodes of laughter and crying, known as pseudobulbar palsy

2. Patients with a deep brain stimulator are unable to undergo brain MRI:
B. False – although the MRI may have artifact and the DBS will need to be reset after the study

3. A deep brain stimulator causes an irreversible brain injury:
B. False – the DBS is adjustable and can even be turned off

4. The major risk of placing a deep brain stimulator is:
A. Infection

Case 5 – Movement Disorders

A 51 year old right handed man has alcoholic cirrhosis of the liver. After experiencing several life threatening gastrointestinal variceal hemorrhages he underwent transcutaneous intrahepatic portosystemic shunting (TIPS) with interventional radiology. Although his variceal bleeding stopped, the patient continued to have a cognitive decline. He was also noted to develop tremors, myoclonic jerks, and significant rigidity in both arms and slowed gait. A brain MRI was obtained.

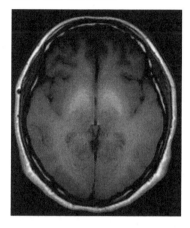

◄ *T1W non-enhanced brain MRI of a patient with Parkinsonism.*

1. There is abnormal T1W signal on the MRI shown above – which part of the brain does it affect?
A. Thalamus
B. Basal ganglia
C. Substantia nigra
D. Motor cortex

2. Which of the following is a hereditary disease of liver degeneration which results in Parkinsonism?
A. Lewy Body dementia
B. Non-alcoholic steatohepatitis
C. Wilson disease
D. Huntington disease

3. Which of the following is a common neurologic sequelae of TIPS?
A. Worsening of hepatic encephalopathy
B. New onset epilepsy
C. Increased risk of ischemic strokes
D. High risk for CNS vasculitis

Case 5 – Hepatocerebral Degeneration

There are many neurological consequences of chronic liver failure. The most common is hepatic encephalopathy (HE) – importantly, in advanced HE the serum ammonia levels do not correlate well with the level of cognitive impairment. HE worsens after TIPS because even less serum ammonia is removed by the liver.

Movement disorders, such as Parkinsonism, dystonia, myoclonus and tremors are very common in advanced liver disease. One explanation is the accumulation of manganese within the basal ganglia. A similar situation occurs in Wilson disease, in which a hereditary error in copper metabolism results in excess copper deposition in the brain, eye and in the liver.

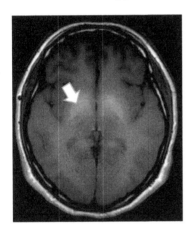

◄ *T1W non-enhanced brain MRI, showing manganese accumulation in the basal ganglia (arrow). This is commonly seen after portosystemic shunting or in patients exposed to high levels of manganese (i.e. welders). A similar picture can be seen in Wilson disease, in which copper is deposited in the basal ganglia.*

1. There is abnormal T1W signal in which part of the brain on the MRI shown above? **B. Basal ganglia**

2. Which of the following is a hereditary disease of liver degeneration which results in Parkinsonism? **C. Wilson disease, which causes brain copper accumulation**

3. Which of the following is a common neurologic sequelae of TIPS? **A. Worsening of hepatic encephalopathy, as blood is shunted through the liver and even less ammonia and other toxins are removed than before TIPS**

▶ *An interventional radiologist performing a TIPS procedure.*

Case 6 – Movement Disorders

A 65 year old right handed woman is referred to you for suspected Parkinsonism. She first developed symptoms of tremor one year ago. Since that time, she has rapidly progressed to significant bilateral rigidity. Because of her frequent and severe falls she can no longer walk unassisted. Often the falls appear to happen "out of the blue" without provocation.

When you examine her you note that her head is tilted back and she has a surprised look on her face. She can track your finger side to side on exam, but not up and down. Her family is at their wits end from the rapid progression of her symptoms, who note that she has lost her usual pleasant demeanor.

The brain MRI shows only non-specific atrophy of the midbrain.

1. Which of the following are atypical features for idiopathic Parkinson disease in this case?
A. Rapid disease progression
B. Complete inability to look up or down
C. Frequent falls, early in the disease course
D. Neck hyperextension
E. All of the above
F. None of the above

2. Should this patient be trialed on carbidopa/levodopa for the Parkinsonism symptoms (rigidity, tremor, bradykinesia)?
A. Yes – it is often helpful in these cases
B. Yes – although it is rarely helpful in these cases
C. No – it can precipitate encephalopathy
D. No – because it is never helpful in these cases

Case 6 – Progressive Supranuclear Palsy

This patient has Progressive Supranuclear Palsy, one of several so-called "Parkinsons Plus" syndromes. These rare conditions are characterized by Parkinson disease-like features such as rigidity, bradykinesia, tremor and falls, but are typically more rapidly progressive, and with additional features depending on the specific disease type. A trial of carbidopa/levodopa is often warranted, as some patients may see mild improvement in motor symptoms. The focus is on excluding treatable mimics (i.e. Wilson disease or encephalitis) and palliation of symptoms. In most cases, you need only be aware that these rare conditions exist and the red flags that separate them from Parkinson disease. Note that the diagnosis of these conditions is exceedingly difficult, and often occurs on post-mortem analysis.

Parkinson Disease Red Flags
- Upper motor neuron signs (increased reflexes, spasticity)
- Autonomic instability (orthostatic hypotension, etc.)
- Severe vertical gaze impairment
- Early hallucinations and dementia
- Rapid progression (severe symptoms in 1 – 3 years)
- Cortical signs (agnosia, apraxia, etc.)
- Myoclonus

1. Which of the following are atypical features for idiopathic Parkinson disease in this case? **E. All of the above**

2. Should this patient be trialed on carbidopa/levodopa for the Parkinsonism?
B. Yes – although it is rarely helpful in cases of Parkinsons plus, it's worth a short trial of medication in case the patient has symptomatic benefit

Parkinsons Plus Syndromes:

Progressive Supranuclear Palsy
- Frequent falls
- Impairment of vertical gaze
- Surprised look due to eyelid apraxia

Corticobasal Degeneration
- Asymmetric extreme rigidity, apraxia or clumsiness
- Cognitive impairment

Multi-System Atrophy
- Prominent autonomic features, such as orthostatic hypotension
- Ataxia
- Upper motor neuron signs

Other Causes of Movement Disorders

There are many other causes of movement disorders, many of them rare. Here are a few illustrative examples.

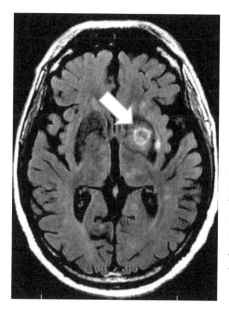

◄ This patient presented with right sided **hemiballismus**.

Hemiballismus is a very chaotic, hyperactive movement disorder, with wild and uncontrollable flinging of the arms and legs. It is often due to a focal lesion of the subthalamic nucleus (recall that this is where DBS leads are placed to increase movement in Parkinson patients). The patient here had HIV and had a toxoplasmosis abscess in the left subthalamic nuclei (arrow). Another common cause is stroke involving the same area.

Myoclonus is a very rapid, short duration muscle jerk, which can look like a twitch. Many people experience this when falling asleep (called a hypnic jerk). However, it can occur pathologically as well. There are many causes, including focal seizures, corticobasal degeneration and hepatocerebral degeneration – but the most common cause is probably medications.

Some common medications that can cause myoclonic jerks:
- Antidepressants (SSRI's, TCA's and MAO-I's)
- Levodopa
- Gabapentin (in renal failure)
- Lithium
- Some antibiotics, such as quinolones or cephalosporines
- Opiates, especially at high doses or in withdrawal

Chorea is an involuntary, writhing movement. It comes from the Greek word for dance, because it's a continuous, smooth and often sinusoidal type movement. Although the classic condition causing chorea is Huntington disease, there are other causes of chorea. One relatively common etiology is hyperglycemia – the mechanism for this is unclear, however. A feature of chorea is motor impersistence, the inability to sustain a movement. Examples include frequently changing posture or maintain a steady hand grip.

Dystonia is a sustained muscle contraction, for example a writer's cramp.

Akathisia refers to the feeling of needing to move, due to restlessness. It is most often seen as the result of antipsychotic medication, and can lead to repetitive movements. It can also be seen in Parkinson disease, or in drug withdrawal.

Stereotypy refers to the repetitive movement of complex tasks, such as hand waving, foot tapping or head shaking. Most commonly seen in children, including normal children. However, it is also very common in children with autism or developmental delay.

Tardive dyskinesia is a disorder of frequent abnormal movements, usually writhing, fidgety or choreaform movements involving the mouth, tongue or fingers. Tardive dyskinesia typically occurs following the use of dopamine receptor blocking medications, principally antipsychotic medications, but also including antiemetics such as metoclopramide. Even small exposure to dopamine receptor blocking mediations can result in tardive dyskinesia, although the risk increases with increased exposure to the offending medications. Less potent antipsychotic medications, such as second generation medications like olanzapine and quetiapine, have lower rates of tardive dyskinesia. This movement disorder may be permanent.

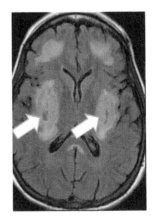

▶ *T2W FLAIR brain MRI, showing a patient with bilateral necrosis of the putamen (arrows) following methanol ingestion, resulting in acquired Parkinsonism. Such acquired Parkinsonism is resistant to treatment with carbidopa/levodopa. Other insults to the basal ganglia, especially cerebrovascular disease (i.e. stroke), can cause acquired Parkinsonism or chorea.*

Case 7 – Movement Disorders

A 40 year old right handed man is evaluated for several years of progressive memory trouble and confusion. His wife notes that over the past few years he has become depressed, has angry outbursts and often acts "like he no longer cares." On exam he is fidgety, and cannot maintain a constant hand grip or stick his tongue out for more than a few seconds. He appears to be constantly in motion, even when sitting in a chair. The patient himself does not seem to be bothered by the movements. The patient's spouse notes that her husband's father had developed similar symptoms several years prior committing suicide.

1. Would you obtain a brain MRI for this patient?
A. Yes   B. No

2. If you did obtain a brain MRI, in which part of the brain might you suspect to see pathology?
A. Cerebellum
B. Basal ganglia
C. Thalamus
D. Hippocampus

3. How would you classify this patient's movements based on the above description?
A. Chorea
B. Myoclonus
C. Dystonia
D. Ataxia
E. Akathisia
F. Hemiballismus
G. Rigidity

4. This patient's father seems to have had the same disease. In autosomal dominant genetic conditions, what are the odds that an affected patient will pass the disease on to their offspring?
A. 1 in 2
B. 1 in 4
C. 1 in 66
D. It is passed on in all cases

Case 7 – Huntington Disease

This young man has slowly progressive cognitive and behavioral problems, as well as progressively abnormal movements most consistent with chorea. A positive family history for a similar disorder is highly suggestive of Huntington disease (HD). HD is a slowly progressive neurodegenerative disease. It is a genetic condition due to CAG trinucleotide repeats with autosomal dominance, meaning children of affected individuals have a 50% chance of inheriting the disease. In fact, children of affected fathers (as compared to affected mothers) have a chance of inheriting a more severe form of the disease – a phenomenon known as anticipation.

1. Would you obtain a brain MRI for this patient?
A. Yes – new onset chorea is concerning for a basal ganglia lesion

2. If you did obtain a brain MRI, where might you suspect to see pathology?
B. Basal ganglia – specifically the caudate nucleus of the basal ganglia

3. How would you classify this patient's movements based on the above description?
A. Chorea – constant, writhing, fidgety movement with motor impersistence

4. This patient's father seems to have had the same disease. In autosomal dominant genetic conditions, what are the odds that an affected patient will pass the disease on to their offspring?
A. 1 in 2

> **Teaching Point:** Rates of depression and suicide are extremely high in Huntington disease, approaching 8%.

▲ *T2W brain MRI, showing normal caudate nuclei on the left (arrow) and atrophied caudate head (arrowhead) on the right, typical of Huntington disease.*

Case 7 – Movement Disorders

You are called to the emergency department to see a 23 year old right handed woman who had acute onset of bilateral upward gaze and facial pain. When you arrive her neck is extended and she is looking up at the ceiling. Her jaw is clenched and she appears to be in severe pain. She had presented to the emergency department several hours ago for intense nausea and vomiting and had received IV saline with potassium, IM Ketorolac, PO thiamine, IV prochlorperazine and IV magnesium.

1. Does this patient require urgent brain imaging?
A. Yes – stat head CT
B. Yes – stat brain MRI
C. Yes – CT and CT angiogram
D. No brain imaging is necessary

2. Which of the following is the most likely cause of the patient's symptoms?
A. IV saline with potassium
B. IM ketorolac (Toradol)
C. PO thiamine
D. IV prochlorperazine (Compro)
E. IV magnesium
F. None of the above – this is a sporadic condition

3. This is an example of which of the following types of movement disorders?
A. Chorea
B. Myoclonus
C. Dystonia
D. Ataxia
E. Akathisia
F. Hemiballismus
G. Rigidity

4. Which of the following is the best immediate treatment for this patient's condition?
A. IV tPA
B. IV Dantrolene
C. IV Diphenhydramine
D. IM Olanzapine

Love movement disorders? Grab a copy of *Movement Disorders in Clinical Practice* by K. Ray Chaudhuri and William G. Ondo.

Case 7 – Acute Dystonic Crisis

This young woman is experiencing an oculogyric crisis – a form of acute dystonic reaction. Recall that a dystonia is an abnormal, sustained muscle contraction. Common examples include torticollis, a painful twisting of the neck, or writer's dystonia, also called a writer's cramp. These can arise sporadically, but can also be induced by acute use of a dopamine receptor blocking medication. Young adults seem to be more prone to developing an acute dystonic crises, and anti-emetics such as prochlorperazine or metoclopramide are common culprits. Treatment is typically with either a benzodiazepine, such as low dose diazepam, or an anticholinergic agent such as diphenhydramine.

1. Does this patient require urgent brain imaging?
D. No brain imaging is necessary

2. Which of the following is the most likely cause of the patient's symptoms? **D. IV prochlorperazine (Compro), because of its dopamine antagonist activity**

3. This is an example of which of the following types of movement disorders?
G. Dystonia

4. Which of the following is the best immediate treatment for this patient's condition?
C. IV Diphenhydramine

Acute Dystonic Reactions

Oculogyric Crisis: Spasm of the extraocular muscles, causing upward gaze deviation – may have neck and jaw involvement.

Torticollis: Head flexed or turned to the side. When the neck and back are both painfully extended this is called **opisthotonos**.

Laryngospasm: Spasm of the vocal cords, making it difficult to speak or breath.

Blepharospasm: Abnormal contraction of the eyelids.

Buccolingual Crisis: Dysarthria, grimacing and forced smile (called a *risus sardonicus*).

Case 8 – Movement Disorders

A 62 year old right handed woman had to quit her job as a sales executive because she was becoming increasingly forgetful and disorganized. Her husband notes that she had become increasingly forgetful and easily lost, which was very unlike her. Increasingly, she reported seeing small animals or bugs in the house, mostly in the evening, although her husband states there was nothing to be seen. He also notes that some days she will be lethargic and confused, while other days she appears much closer to her baseline. The following year, she developed a left sided resting tremor, prompting her to visit you. On exam, she has left greater than right cogwheel rigidity, a mildly masked facial expression and is very slow with rapid alternating movements.

1. This patient's presentation is most concerning for which disease?
A. Lewy Body dementia
B. Non-alcoholic steatohepatitis
C. Wilson disease
D. Huntington disease
E. Parkinson disease with dementia
F. Alzheimer disease
G. Frontotemporal dementia
H. Substance abuse disorder

2. Which of the following is not a common feature of this condition?
A. Delusions
B. Orthostatic hypotension
C. Auditory and visual hallucinations
D. Manic episodes
E. Depression
F. Apathy

3. On exam this patient has overt Parkinsonism. Patients with this condition typically react very poorly to which medication class?
A. Antipsychotics
B. NSAIDs
C. Carbapenem antibiotics
D. Antifungals
E. MAO-B inhibitors

Case 8 – Lewy Body Dementia

This patient has Lewy Body Dementia (LBD), a disease which in many ways is similar to Parkinson Disease. However, in LBD dementia and hallucinations present early, followed by the parkinsonian movement features. A significant number of patients with Parkinson Disease also develop dementia at some point in their illness, usually after several years. A key distinguishing feature is that significant cognitive dysfunction precedes the motor dysfunction in LBD. The underlying pathology is thought to be very similar in both Parkinson and LBD.

Patients often report a history of REM sleep behavior disorder, commonly acting out their dreams. They develop hallucinations in dim light, most often seeing animals or bugs, which they often find distressing. Many will have cognitive fluctuations, characterized by periods of blanking out or having confused and bizarre behavior, only to improve spontaneously.

Lastly, these patients are often exquisitely sensitive to antipsychotic medications, which can profoundly worsen their parkinsonism or cause confusion.

1. This patient's presentation is most concerning for which disease?
A. Lewy Body dementia

2. Which of the following is <u>not</u> a common feature of this condition?
D. Manic episodes – delusions, depression, apathy, autonomic dysfunction and both auditory and visual hallucinations are common in LBD. REM sleep behavior disorders are also a frequent, early feature.

3. On exam this patient has overt Parkinsonism. Patients with this condition typically react very poorly to which medication class?
A. Antipsychotics – patients with LBD will respond poorly to any dopamine antagonist medication, including many of the anti-emetics. It may significantly worsen their Parkinsonism. Very low doses of quetiapine are sometimes used for patient safety if the hallucinations and delusions are severe.

Notes

Paul D. Johnson, MD

Neuromuscular Disorders

"Look at the bright side, nobody got hurt."
- "People got hurt."
- "I'm saying I think they died quickly so I don't think that they got hurt."
The Nice Guys

A Short Introduction to Electromyography and Nerve Conduction Studies

The field of neuromuscular neurology is concerned with the control of the body's skeletal muscle, from the corticospinal tract, through the spinal cord, to peripheral motor nerves, the connection between motor nerves and muscle (called the neuromuscular junction) and even the health and function of the muscle itself.

It turns out that muscles requires a functioning connection to a motor nerve in order to remain alive and active. While strokes and other brain injuries can cause weakness, they typically leave the motor nerve (called lower motor neurons, connecting the spinal cord to the muscle), intact. However, injury to the lower motor neuron or muscle — collectively known as a "motor unit" (see Figure 1) can lead to weakness, atrophy and loss of reflexes.

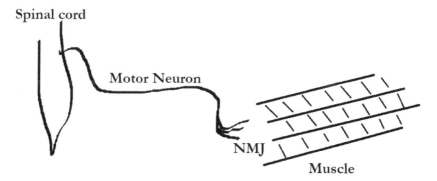

▲ Figure 1. *Diagram of the motor unit, showing a motor neuron running from the anterior horn of the spinal cord to the muscle. Each motor neuron innervates a specific set of muscle fibers. The peripheral motor nerve is a lower motor neuron; the motor nerve running from the brain to the spinal cord is the upper motor neuron.*

This is a good time to point out that the peripheral nerves, like the motor or sensory neurons, are composed of an axon – the cell body – and an outer covering called myelin (Figure 2). The myelin acts like insulation on a wire, dramatically improving the transmission of nerve signals.

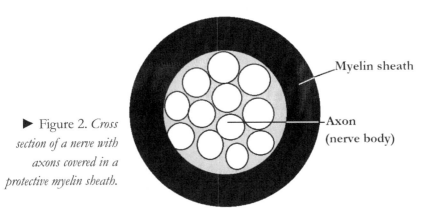

▶ Figure 2. *Cross section of a nerve with axons covered in a protective myelin sheath.*

Nerve Conduction Study (NCS) is an electrical test that certain neurologists and most rehab medicine doctors can do to measure the health of the axons and the myelin sheath. We won't get into detail here, but the EMG test can give a quantitative measurement of nerve function, including the speed of transmission and volume of electrical signal getting through. This can be measured at different points along the length of the nerve. Thus, you can compare proximal or distal nerve function, compare side to side or compare nerves in different limbs. This is extremely helpful for measuring and diagnosis diseases of the peripheral nerves.

Electromyography (EMG) is an electrical test which evaluates the health of the muscle. It is often done with NCS. The examiner places a small needle into muscle which measures the muscle's electrical properties – this actually tells us a lot about the health of the muscle and its relationship to the nerve. It can help differentiate between intrinsic muscle diseases, such as myopathy, and diseases of the nerve leading to secondary atrophy of the muscle. Not surprisingly, both EMG and NCS can be uncomfortable for patients!

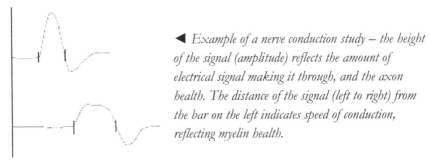

◀ *Example of a nerve conduction study – the height of the signal (amplitude) reflects the amount of electrical signal making it through, and the axon health. The distance of the signal (left to right) from the bar on the left indicates speed of conduction, reflecting myelin health.*

Case 1 – Neuromuscular Disorders

A 34 year old left handed woman with a remote history of hashimoto thyroiditis complains of episodic double vision and droopy eyelids, that becomes progressively worse throughout the day. These symptoms have been worsening over the past few weeks, but in the last few days she has also noted difficulty with swallowing. Her symptoms improve temporarily with rest. On exam you note right greater than left diplopia and slightly disconjugate gaze, with an inability to fully abduct the left eye. She has mild bilateral facial weakness, weakness of the lips, and the longer you speak to her the more nasal her voice becomes.

1. A disorder in which of the following locations is most likely responsible for this young woman's problems?
A. Spinal cord
B. Peripheral nerve axon
C. Peripheral nerve myelin
D. Neuromuscular junction
E. Muscle
F. Motor cortex

2. The condition can be confirmed through which test?
A. Lumbar puncture
B. Serum antibodies
C. Brain MRI
D. Adrenal stimulation testing

3. Patient's with this condition should undergo which of the following additional studies?
A. Chest CT
B. Abdominal CT
C. Pelvic ultrasound
D. CT myelogram

4. Administration of which of the following medications could make this patient's condition dramatically <u>worse</u>?
A. IVIG
B. Rituximab
C. Pyridostigmine (Mestanon)
D. Moxifloxacin

Case 1 – Myasthenia Gravis

Myasthenia Gravis (MG) is the quintessential disease of the neuromuscular junction. It is an autoimmune condition that often affects young adults. It is closely associated with disease of the thymus – thus everyone with a new diagnosis of myasthenia gravis should have a screening contrast enhanced chest CT to evaluate for thymoma. Surgical removal of the thymus appears to be very beneficial for these patients.

The disease is due to antibodies in the neuromuscular junction, which block acetylcholine, the neurotransmitter released by motor nerves. A key features of MG is fatigable weakness, meaning patients get weaker with effort and improve temporarily after rest.

Most patients with MG will have serum antibodies – typically **anti-acetylcholine receptor antibodies**, and less commonly **anti-muscle tyrosine kinase antibodies** (anti-MUSK). Fifteen percent of patients with MG will be antibody negative. In that case, the diagnosis can be supported by single fiber or repetitive nerve stimulation testing on nerve conduction studies.

Certain medications, especially some antibiotics, can make MG much worse and should be avoided – consult with your pharmacist.

Some medications that worsen MG (may be others as well):
- **Antibiotics:** Most aminoglycosides and fluoroquinolones (i.e. ciprofloxacin), azithromycin, clindamycin and sulfonamides
- **Magnesium containing medications**
- **Neuromuscular junction blocking agents**
- **Potentially beta blockers, verapamil, barbiturates**

1. A disorder in which of the following locations is most likely responsible for this young woman's problems? **D. Neuromuscular junction – fatigable weakness is typical**

2. The condition can be confirmed through which test?
B. Serum antibodies (anti-acetylcholine receptor Ab's or anti-MUSK Ab's)

3. Patient's with this condition should undergo which of the following additional studies?
A. Chest CT with contrast to evaluate for thymoma

4. Administration of which of the following medications could make this patient's condition dramatically worse?
D. Moxifloxacin – the other choices are MG treatments; there is a long list of medications that worsen MG, the University of Michigan has a nice list online

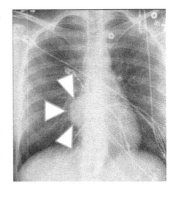

▶ *Chest radiograph showing a very large thymoma adjacent to the heart (outlined with arrowheads). Removing the thymoma leads to improved outcomes, but is not curative.*

Case 2 – Neuromuscular Disorders

A 21 year old right handed man is seen in the office for bilateral foot drop. He has a history of falls as a child, and states he was "never great at sports." On exam he has high-arches and hammertoes. Patellar and achilles tendon reflexes are absent. He has loss of sensation to pin prick in the feet, which gradually improves as you move up the lower leg. On EMG/NCS conduction velocities are uniformly slow, without evidence of conduction block. His blood work is all normal, and he has no other medical issues. His father had similar issues but was never evaluated. (Hint: conduction block is severe, focal slowing in motor nerve transmission, usually due to patchy demyelination of the nerve).

1. A disorder in which of the following locations is most likely responsible for this young man's problems?
A. Spinal cord
B. Peripheral nerve axon
C. Peripheral nerve myelin
D. Neuromuscular junction
E. Muscle
F. Motor cortex

2. Which study would be best for showing the decreased conduction velocity described in the case?
A. EMG only
B. NCS only
C. Both EMG and NCS would show slowed conduction
D. Neither – it's a clinical diagnosis

3. Treatment for his progressive weakness and foot drop should include all of the following, <u>except</u> which?
A. Physical therapy
B. Ankle Foot Orthotic (AFO) use
C. Treatment with cyclophosphamide
D. Screening for diabetes

Do you know the name of this patient's condition? Hint – it's hereditary.

High arches and hammertoes, from muscle atrophy. ▶

Case 2 – Charcot-Marie-Tooth Disease

This young man has Charcot-Marie-Tooth disease (CMT). CMT is a hereditary condition in which the peripheral nerves progressively become dysfunctional, leading to distal weakness, atrophy and sensory loss. It is also called hereditary motor sensory neuropathy (HMSN). There are several types, with different hereditary patterns.

CMT1A is the most common type of CMT, and is autosomal dominant – therefore it tends to be prominent in certain families. The development of bilateral foot drop is common. On exam there will be atrophy of intrinsic muscles of the hand and feet. The result of this muscle atrophy is the characteristic high arches and hammertoes of the feet.

The diagnosis can be confirmed through genetic testing and nerve conduction studies. Treatment is supportive, through physical therapy and use of orthotic devices. In addition, great care should be taken to avoid neurotoxic medications and neurotoxic conditions such as diabetes mellitus.

1. A disorder in which of the following locations is most likely responsible for this young man's problems? **C. Peripheral nerve myelin. Recall that myelin is responsible for the speed of signal transmission, or conduction velocity. Peripheral nerve dysfunction also gives the characteristic 'length dependent' symptoms that he displays**

2. Which study would be best for showing the decreased conduction velocity described in the case?
B. NCS only - recall that EMG is a test of the muscle health

3. Treatment for his progressive weakness and foot drop should include all of the following, except which?
C. Treatment with cyclophosphamide

CMT	Inheritance	Neuropathy Type
Type 1A	Autosomal Dominant	Demyelinating
Type 2	Autosomal Recessive	Axonal
Type 3	Autosomal Recessive	Demyelinating

Case 3 – Neuromuscular Disorders

A 60 year old right handed man complains of weakness and fatigue. He is found to have asymmetric atrophy, isolated to his quadriceps, anterior tibialis, finger flexors and wrist flexors. He complains of mild dysphagia. These symptoms have all been slowly progressive over the past year or more. His serum CK is mildly elevated at 3,000 and his affected muscles are slightly tender.

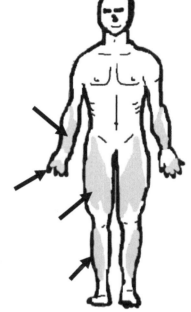

▶ *Areas of weakness and atrophy for this patient are marked in grey, including the finger flexors, forearm flexors, hip flexors and anterior tibialis muscles, although the degree of atrophy is not equal from side to side.*

1. A disorder affecting which of the following locations is most likely responsible for this man's weakness?
A. Spinal cord
B. Peripheral nerve axon
C. Peripheral nerve myelin
D. Neuromuscular junction
E. Muscle
F. Motor cortex

2. Should this patient undergo a muscle biopsy?
A. Yes   B. No   C. Only if he gets worse

3. Which study would be best for showing abnormalities in this case?
A. EMG only   C. Both EMG and NCS
B. NCS only   D. Neither would show any abnormality

Case 3 – Inclusion Body Myositis

Inclusion Body Myositis (IBM) is an example of a myopathy – intrinsic disease of the muscle. IBM is the most common myopathy and occurs sporadically in people over age 50. This patient has the typical pattern of weakness, including slight dysphagia and mildly increased CK (usually not more than 10x greater than normal). Unfortunately, there is no cure for IBM.

The elevated CK and slight muscle tenderness, as well as atrophy are all good indications that the patient has a myopathy. EMG can help to confirm the diagnosis.

Tip: Autoimmune conditions can also cause myopathy. Look for rash, especially on the face, knuckles or hands, to suggest **dermatomyositis.** Myopathy restricted to the proximal muscles may be **polymyositis.** These conditions are responsive to steroids. Muscle biopsy may be helpful in making the diagnosis.

1. A disorder in which of the following locations is most likely responsible for this man's weakness? **E. Muscle – given the diffuse pattern, tenderness and atrophy**

2. Should this patient undergo a muscle biopsy?
A. Yes – muscle biopsy should at least be considered to confirm the diagnosis

3. Which study would be best for showing abnormalities in this case? **A. EMG only, since EMG reflects muscle health, and the nerves are not affected in myopathy**

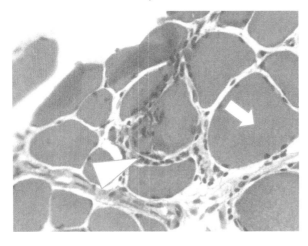

◄ *Example of a muscle biopsy showing muscle fibers (arrow) and inflammatory cells (arrowhead). In IBM you typically see 'inclusions' within the muscle fibers.*

Case 4 – Neuromuscular Disorders

A 24 year old right handed female graduate student is recovering from a recent diarrheal illness when she develops tingling in her fingers and toes, followed by low back pain. Over the next three days she notes progressive difficulty walking, followed by difficulty going up stairs. On exam her sensation is intact and there is no muscle tenderness or atrophy. Her achilles and patellar reflexes are absent, although her biceps reflexes are normal.

1. A disorder affecting which of the following locations is most likely responsible for this young woman's problems?
A. Spinal cord
B. Peripheral nerve axon
C. Peripheral nerve myelin
D. Neuromuscular junction
E. Muscle
F. Motor cortex

2. All of the following are possible complications of this patient's condition except for which?
A. Malignant hyperthermia
B. Autonomic instability
C. Respiratory failure
D. Chronic residual weakness

3. Which of the following is not a typical finding in this condition?
A. Mildly increased CSF protein
B. Inflammation on muscle biopsy
C. Enhancement of nerve roots on MRI
D. Conduction block on NCS

4. Which of the following is the most effective initial treatment for this condition?
A. Oral prednisone
B. IV prednisone
C. IVIG
D. Azathioprine

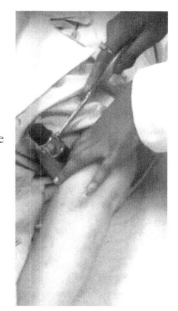

▶ *The patient has rapidly ascending weakness and loss of reflexes.*

Case 4 – Guillan-Barré Syndrome

This woman has Guillan-Barré syndrome (GBS), also known as Acute Inflammatory Demyelinating Polyradiculoneuropathy (AIDP). This is an immune mediated inflammatory condition affecting the spinal nerve roots. It is frequently triggered by infection, especially gastrointestinal or respiratory infections, but also potentially by the influenza vaccine.

Classically, patients develop tingling in the extremities, followed several days later by ascending weakness and neuropathic low back pain. Depending on the severity of the attack, the weakness may involve the respiratory muscles. Weakness ranges from mild to complete quadriplegia. Autonomic dysfunction, especially sinus tachycardia, is common and patients should be monitored on telemetry.

Typical first line therapy is 2 grams/kg of IVIG, given in divided doses over 2 to 5 days. However, in severe cases therapeutic plasma exchange is also beneficial, and may be better than IVIG.

CSF usually shows elevated protein without an increase in inflammatory cells, called **cytoalbuminologic dissociation**. The EMG/NCS can be diagnostic, but should be delayed at least 2 weeks from symptom onset as early studies can be falsely negative. You may see spinal nerve root enhancement on MRI.

> **Tip**: There are many variants of AIDP, which can involve the cranial nerves or extra-ocular muscles. A common variant is Miller Fischer syndrome, characterized by ataxia, ophtalmoplegia and areflexia but minimal weakness. The chronic form (CIDP), lasts > 4-6 weeks and may require maintenance therapy with IVIG.

1. A disorder in which of the following locations is most likely responsible for this young woman's problems? **C. Peripheral nerve myelin – specifically in the nerve roots**

2. All of the following are possible complications of this condition except for which? **A. Malignant hyperthermia – autonomic failure, potential respiratory weakness and chronic weakness are all possible**

3. Which of the following is not a typical finding in this condition? **B. Inflammation on muscle biopsy – that would suggest a myopathy**

4. Which of the following is the most effective initial treatment for this condition? **C. IVIG – in severe cases plasma exchange may be better**

Case 5 – Neuromuscular Disorders

A 65 year old man develops 9 months of progressive weakness in his left hand, followed by left arm weakness and then difficulty swallowing. On exam there is weakness and significant muscle atrophy primarily in the left arm, but also in his right leg. The left arm is relatively flaccid with reduced reflexes. There is an extensor plantar response on the right, as well as sustained clonus in that ankle. The patient has mild dysarthria. He denies pain other than occasional muscle cramps. His wife has also noted occasional rippling movements across his muscles. On your exam you notice similar muscle ripples and atrophy in the tongue.

1. This patient's condition is most likely due to which type of abnormality?
A. Lower motor neuron disease
B. Upper motor neuron disease
C. Both upper and lower motor neuron disease
D. Toxin induced myopathy

2. Which of the following is the next best diagnostic step?
A. Brain MRI
B. EMG/NCS
C. Lumbar puncture
D. Muscle biopsy
E. Nerve biopsy

3. Which of the following findings suggests that a cervical spinal cord injury is not the cause of the patient's symptoms?
A. The presence of tongue atrophy and fasiculations
B. The left arm atrophy and weakness
C. The right leg atrophy
D. The right leg extensor plantar response and ankle clonus

Case 5 – Amyotrophic Lateral Sclerosis

This patient most likely has ALS. Consider ALS whenever there is focal onset of weakness, especially with muscle wasting and brisk reflexes, that spreads and progresses over time.

ALS is a disease of both upper motor neurons (cortical motor neurons) and lower motor neurons (peripheral nerves originating in the spinal cord). Making the diagnosis requires finding evidence of both upper and lower motor neuron dysfunction in the same body region, affecting at least 3 different regions. It is a slowly progressive condition, currently with no known cure.

The EMG/NCS is crucial to making the diagnosis of ALS. This study can identify subtle findings of motor neuron dysfunction before it becomes clinically apparent.

Upper Motor Neuron Signs
- Increased tone (spasticity)
- Brisk reflexes (i.e. clonus or extensor plantar response)
- Minimal muscle wasting

Lower Motor Neuron Signs
- Muscle wasting (atrophy)
- Reduced or absent reflexes
- Fasiculations
- Decreased muscle tone (flaccid tone)

Riluzole is an expensive oral medication that can delay progression of ALS by several months. Otherwise, treatment is supportive for the many complications of ALS, including cramps, falls, choking, respiratory failure, constipation, depression, fatigue and more.

1. This patient's condition is most likely due to which type of abnormality?
C. Both upper and lower motor neuron disease – however, rare variants of ALS do exist with only isolated upper or lower motor neuron features

2. Which of the following is the next best diagnostic step?
B. EMG/NCS – muscle biopsy is not needed to diagnose ALS

3. Which of the following findings suggests that a cervical spinal cord injury is not the cause of the patient's symptoms?
A. The presence of tongue atrophy and fasiculations – bulbar (brainstem) symptoms strongly argue that this is not due to a cervical myelopathy.

Case 6 – Neuromuscular Disorders

A 4 year old boy is brought in for an evaluation of falls. He has diffuse lower extremity weakness and walks on his toes. On exam he is noted to have an exaggerated arch to his lower back as well as very large, firm calves. When he moves from a sitting to standing position he uses his hands to push himself off the floor and then he pushes his upper body up by placing his hands on his thighs. His parents suspected that his IQ is somewhat lower than that of his siblings.

▶ *This young patient was noted to have very thin thighs, but very large and firm calves. His shoulders were also atrophied.*

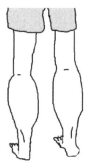

1. A disorder affecting which of the following locations is most likely responsible for this young man's problems?
A. Spinal cord
B. Peripheral nerve axon
C. Peripheral nerve myelin
D. Neuromuscular junction
E. Muscle
F. Motor cortex

2. Which of the following is the next best diagnostic step?
A. Genetic testing
B. Muscle biopsy
C. Lumbar puncture
D. Nerve conduction studies
E. Nerve biopsy

3. Would you expect the serum CK to be: _____
A. Slightly elevated
B. Significantly elevated
C. Normal
D. Slightly decreased
E. Significantly decreased

Case 6 – Duchenne Muscular Dystrophy

This is a typical case of Duchenne muscular dystrophy. Muscular dystrophies are an inherited group of muscle disease with progressive muscle weakness. The most common conditions are Duchenne and Becker muscular dystrophy – both are X-linked genetic conditions, and therefore almost always affects boys. The genetic mutation results in dysfunction of the muscle proteins, leading to gradual breakdown of the muscle and replacement with fatty tissue. Due to this breakdown, the serum CK is highly elevated, up to 100x normal.

Boys with Duchenne muscular dystrophy tend to present with weakness between ages 3 and 5, and are typically in a wheelchair by their early teenage years. Becker muscular dystrophy is a more mild disease, and presents at a later age.

Cardiomyopathy is extremely common in Duchenne and Becker muscular dystrophy, as well as other myopathies, and is a significant source of morbidity and mortality. Cardiac screening is an important part of the evaluation and management of all patients with myopathy.

Teaching Point: Although the muscular dystrophies are not inflammatory conditions (like IBM or polymyositis), prednisone delays disease progression.

1. A disorder in which of the following locations is most likely responsible for this young man's problems?
E. Muscle – non-inflammatory muscle dysfunction

2. Which of the following is the next best diagnostic step?
A. Genetic testing – muscle biopsy can also be done, but when definitive genetic testing is available there is no need to perform a painful and invasive procedure

3. Would you expect the serum CK to be: _____
B. Significantly elevated – up to 100x normal

▶ *This is an example of pseudo-hypertrophy (arrow) of the calf. The muscle is replaced by connective tissue..*

Case 7 – Neuromuscular Disorders

A 54 year old woman is seen in the clinic for progressive paresthesias in both feet and ankles. She has no prior medical history, denies tobacco use, drinks socially and has a family history of diabetes and hypertension. Her symptoms have been present for several years, but recently the paresthesias have become painful – she describes a burning sensation in the feet, especially while walking and at night. In addition, she is having some balance problems, having tripped several times when moving around in her room at night. On exam she has severely reduced vibratory sensation in both feet, as well as reduced sensation to temperature. Pin prick sensation is reduced, but still present, on the dorsum of the feet. Achilles reflexes are absent.

1. In order to screen this patient for an autonomic neuropathy in addition to her sensory neuropathy, you might check all of the following except?
A. Orthostatic vital signs
B. History of sexual dysfunction
C. Ulcers in the mouth
D. Early satiety and diarrhea

2. All of the following are typical first line blood tests for peripheral neuropathy except which?
A. Serum lead levels
B. Fasting glucose
C. Serum vitamin B12 level
D. Thyroid Stimulating Hormone (TSH)
E. Erythrocyte Sedimentation Rate (ESR)
F. Serum Immunofixation Electrophoresis (IFE)

3. In length dependent peripheral neuropathy, the hands often become involved _____.
A. At the same time that the feet are affected
B. When the neuropathy reaches the knees
C. When the neuropathy reaches the neck
D. First, before the feet are affected

4. This patient is at higher risk for which other condition, as a direct result of having neuropathy?
A. Seborrheic dermatitis
B. Frontal balding
C. Bruxism
D. Carpal tunnel syndrome

Case 7 – Peripheral Neuropathy

This patient appears to have a length dependent, symmetric peripheral neuropathy. The longer the nerve, the more susceptible it is to injury from toxins (such as alcohol) or from microvascular disease (such as diabetes mellitus). Diabetes is the most common cause of length dependent peripheral neuropathy in the United States. This patient's HbA1c was 8.9%.

Initial Labs for Peripheral Neuropathy
- Serum B12
- Fasting glucose or HbA1c
- Thyroid function
- Serum electrophoresis and Immunofixation
- ESR

There are several types of peripheral nerve: large motor fibers, large sensory fibers (which sense proprioception), small sensory fibers (pain nerves) and autonomic nerves. A disease may affect some or all of the nerve types.

Most peripheral neuropathies are slowly progressive. **Rapid onset within 1 month or less** is suggestive of:

- Toxin/drug exposure (i.e. chemotherapy, arsenic, etc.)
- Guillain-Barré syndrome (may be a sensory variant)
- Infection (HIV, Lyme disease, hepatitis C)
- Vasculitis
- Porphyria

1. In order to screen this patient for an <u>autonomic</u> neuropathy in addition to her sensory neuropathy, you might check all of the following <u>except</u>?
C. Ulcers in the mouth – oral ulcers can be caused by Niacin deficiency

2. All of the following are typical first line blood tests for peripheral neuropathy <u>except</u>?
A. Serum lead levels – labs should be customized based on history and exam

3. In length dependent peripheral neuropathy, the hands often become involved: **B. When the neuropathy reaches the knees, at which point the hands often become involved**

4. This patient is at higher risk for which condition as a <u>direct result</u> of having neuropathy? **D. Carpal tunnel syndrome, and other compressive neuropathies**

Signs of Autonomic Neuropathy
- Orthostatic hypotension
- Erectile dysfunction
- Gastroparesis (early satiety)
- Diarrhea or constipation

Case 7 – Peripheral Neuropathy

You have determined that this patient has a symmetric, length dependent peripheral neuropathy which is most likely due to poorly controlled diabetes mellitus. An EMG/NCS was obtained which showed length dependent, axonal motor and sensory neuropathy, consistent with this diagnosis. The other initial lab evaluation for peripheral neuropathy is unremarkable, other than a serum B12 of 300 pg/ml (normal is > 200 pg/mL).

1. How would you further investigate the serum B12 of 300 pg/mL?
A. No further evaluation – it is within the normal range
B. Retest in 3 months
C. Check serum methylmalonic acid
D. Check serum porphobilinogen deaminase levels

2. Which of the following is <u>not</u> a good medication for treating long term neuropathic pain?
A. Gabapentin
B. Nortriptyline
C. Duloxetine
D. Oxycodone
E. Amitriptyline
F. Pregabalin

3. Which of the following is <u>not</u> a red flag for a condition other than a diabetic peripheral neuropathy?
A. Onset of symptoms over 4 weeks
B. Prominent ataxia
C. Concurrent development of facial nerve palsy
D. Early development of proximal weakness
E. Development of ulcers on the feet
F. Concurrent onset of neuropathy in the hands and feet

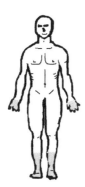

◀ *Classic "stocking-glove" distribution of symmetric, length dependent peripheral neuropathy.*

Case 7 – Peripheral Neuropathy

The patient had an EMG/NCS demonstrating an axonal neuropathy – this is consistent with a toxic/metabolic cause of nerve injury such as diabetes. A demyelinating condition would suggest another process. Note that neuropathic pain medications relieve the pain of peripheral neuropathy, but do not slow loss of sensation or restore sensation. These medications tend to be sedating, and patients should be cautioned that they may need to wait 2-3 months before they see the full benefit of treatment.

> **Medication for Painful Neuropathy:**
> - Tricyclics (nortriptyline, amitriptyline)
> - Venlafaxine or duloxetine
> - Gabapentin or pregabalin
> - Topical lidocaine or capsaicin cream
> - Combinations of above medications
> - Some antiepileptics, such as carbamazepine

1. How would you further investigate the serum B12 of 300 pg/mL?
C. Check serum methylmalonic acid – when B12 is below 400 pg/mL patients can develop high serum levels of methylmalonic acid, which is the pathogenic substance for neuropathy when chronic and can damage the spinal cord, causing ataxia and weakness when acute

2. All of the following are good choices for treating neuropathic pain, with the exception of which? **D. Oxycodone – opiates are a poor choice for neuropathic pain, especially for chronic pain**

3. Which of the following would not be a red flag that this patient has something other than a diabetic peripheral neuropathy?
E. Development of ulcers on the feet – diabetic foot ulcers are common complications of peripheral neuropathy

Teaching point: Small fiber (painful) neuropathies cannot be detected by nerve conduction studies – the nerves are just too small. Only large fiber sensory or motor neuropathies are detectable. However, a skin biopsy may be helpful to identify a small fiber neuropathy.

Case 8 – Neuromuscular Disorders

A 26 year old man presents with rapidly worsening gait – he is clumsy and ataxic, and falls frequently. On exam he has absent proprioceptive and vibratory sensation in his feet, and ataxia on heel-shin exam and finger-to-nose exam. His reflexes are brisk, and he has diffuse weakness. Although he denies drug abuse, his family states that his home is filled with empty nitrous oxide canisters, which he has abused for years.

1. You ask the patient to touch his chin to his chest, which results in a shooting pain down his back. What is this sign called?
A. Lhermitte's sign
B. Spurling's sign
C. Chovstek's sign
D. Collier's sign

2. Which condition does this patient most likely have?
A. Brown-Séquard syndrome
B. Toxic metachromatic leukoencephalopathy
C. Subacute combined degeneration
D. Acquired folate deficiency
E. HTLV-2 infection

3. Which two lab tests will be most helpful to confirm the diagnosis?
A. Methylmalonic acid and folate levels
B. Thiamine and vitamin B12 levels
C. Folate and thiamine levels
D. B12 and methylmalonic acid levels

4. Which anatomical structures are affected in this condition?
A. Dorsal columns and lateral cortical spinal tracts
B. Rubrospinal and lateral cortical spinal tracts
C. Dorsal columns and spinothalamic tracts
D. Spinothalamic and lateral cortical spinal tracts

Case 8 – Subacute combined degeneration

This patient has subacute combined degeneration, in which vitamin B12 deficiency leads to accumulation of serum methylmalonic acid (MMA). In addition to causing peripheral neuropathy, the presence of MMA can result in degeneration of the dorsal columns, leading to ataxia and proprioceptive loss, as well as to cortical spinal tract degeneration which results in weakness. Patients may become encephalopathic as well. Nitrous oxide exposure is a common cause of acquired vitamin B12 deficiency, and is also commonly seen in dentists. Pernicious anemia is another common cause of B12 deficiency, which you can evaluate for by checking anti-intrinsic factor Ab's.

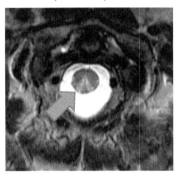

◄ *T2W cervical spine MRI. Note the selective involvement (hyperintensity) of the dorsal columns (arrow).*

1. You ask the patient to touch his chin to his chest, which results in a shooting pain down his back. What is this sign called?
A. Lhermitte's sign – neck flexion causes an 'electric shock' down the spine. In Spurling's downward pressure on the patients rotated head reproduces cervical radiculopathy. Collier's sign is the retraction of the eyelids in dorsal midbrain syndrome, and Chovstek's sign is related to tetany seen in hypocalcemia.

2. Which condition does this patient most likely have?
C. Subacute combined degeneration

3. Which two lab tests will be most helpful to confirm the diagnosis?
D. B12 and methylmalonic acid levels – note that vitamin B12 is a good screening test, and MMA levels are confirmatory

4. Which anatomical structures are affected in this condition?
A. Dorsal columns and lateral cortical spinal tracts

Causes of Myelopathy:

- Vitamin E deficiency
- Copper deficiency – can be provoked by excess zinc intake
- HIV infection – can also affect dorsal columns primarily
- VZV vasculitis of the spinal cord
- HTLV infection
- Vascular malformation – spinal dural AV fistula

Case 9 – Neuromuscular Disorders

An elderly man who had been on both aspirin and high dose rosuvastatin for several years is started on verapamil for hypertension, metformin for diabetes and fenofibrate for hyperlipidemia. Two months later he presents with complaints of diffuse proximal muscle pain, fatigue, and difficulty going up and down stairs. His muscles are slightly tender. His serum CK level is 8,000.

1. All of the following would support a diagnosis of myopathy, except for which?
A. Isolated weakness without sensory loss
B. Brisk reflexes
C. Elevated serum CK
D. Myalgias

2. Which of the following medications is <u>most likely</u> responsible for this patient's symptoms?
A. Aspirin
B. Atorvastatin
C. Verapamil
D. Fenofibrate
E. Metformin

3. Once the offending agent is stopped, symptoms should improve within what time frame?
A. 1 – 3 days
B. 3 – 10 days
C. 1 week to 1 month
D. 1 – 3 months

4. If the patient did not tolerate high dose statin due to myalgias, should he try a lower dose or lower potency statin, or avoid statins altogether?
A. Yes, try a lower dose/ lower potency medication
B. No, all statins should be avoided

Case 9 – Medication Induced Myopathy

This patient has a statin induced myopathy. Statin myopathies are typically dose related and resolve within 1 – 4 weeks of stopping the offending medication. There are many medications that interfere with statin metabolism and can trigger statin induced myopathy (see box).

The typical presentation is proximal, symmetric muscle pain, weakness and fatigue. The hip muscles are frequently affected. Serum CK is typically, but not always, elevated. Withdrawing the offending agent is usually sufficient treatment unless frank rhabdomyolysis is present.

> **Medications that may precipitate statin myopathy:**
>
> - Cyclosporine
> - Calcium channel blockers (like verapamil)
> - Fibrates
> - Many HIV medications
> - Amiodarone
> - Grapefruit juice (inhibits CYP3A4 hepatic metabolism)

1. All of the following would support a diagnosis of myopathy except for which?
B. Brisk reflexes – this is an upper motor neuron sign

2. Which of the following medications is <u>most likely</u> responsible for this patient's symptoms? **B. Rosuvastatin - statins with less hepatic metabolism include pravastatin and atorvastatin, but precipitating medications may also need to be stopped**

3. Once the offending agent is stopped, symptoms should improve within what time frame? **C. 1 week to 1 month**

4. If the patient did not tolerate high dose statin to myalgias. Should he try a lower dose or lower potency statin, or avoid statins altogether?
A. Yes, try a lower dose/ lower potency – many patients will tolerate another medication, such as rosuvastatin or pravastatin

Tip: Rarely, patients will develop a severe, progressive myopathy related to statin use. This is due to the development of anti-HMG-CoA reductase antibodies.

Case 10 – Neuromuscular Disorders

A 63 year old right handed woman complains of several months of progressive weakness in the proximal shoulders and hips. There is no pain, tenderness or muscle atrophy. She does not have sensory changes, nor any bulbar symptoms such as ptosis or dysarthria. With repetitive use of her proximal arms and legs she feels that she temporarily feels stronger. Her medical history is notable only for well controlled hypertension and a long history of tobacco use.

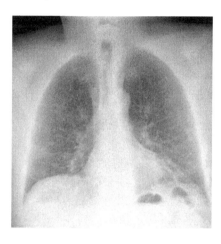

◄ *This patient's chest radiograph appears grossly abnormal.*

1. Which of the following is the most likely cause of this patient's proximal weakness?
A. Myositis
B. Myasthenia gravis
C. Rheumatoid arthritis
D. Lambert-Eaton myasthenic syndrome

2. This patient's condition is most closely associated with which other disorder?
A. Lung adenocarcinoma
B. COPD
C. Esophageal cancer
D. Small cell lung cancer

Case 10 – Lambert-Eaton Myasthenic Syndrome

This patient has Lambert-Eaton Myasthenic Syndrome (LEMS), a disorder in which there is reduced acetylcholine (Ach) release from pre-synaptic nerve terminals, resulting in proximal weakness. With repetitive exercise, calcium concentrations increase, eventually facilitating Ach release and temporarily improving strength. Previously absent reflexes can return temporarily after exercise. LEMS may mimic a myopathy due to the predominantly proximal weakness. Bulbar signs and ophthalmoplegia, common in myasthenia gravis, are rare in LEMS.

In the majority of cases, LEMS is a paraneoplastic phenomenon, most commonly associated with small cell lung cancer. However, this immune mediated condition does occasionally occur in the absence of a neoplasm, in which case patients often have other autoimmune conditions.

The antibody is directed against P/Q-type voltage gated calcium channels.

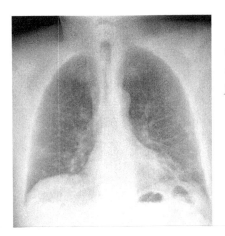

◀ *This chest radiographic demonstrates diffuse interstitial fibrosis, as well as several smaller nodules which were positive for small cell lung cancer.*

1. Which of the following is the most likely cause of this patient's proximal weakness? **D. Lambert-Eaton myasthenic syndrome (LEMS)**

2. This patient's condition is most closely associated with which other disorder? **D. Small cell lung cancer – LEMS is most frequently a paraneoplastic disorder, and may improve after treatment of the underlying tumor**

Neuromuscular Quick Reference

Upper Extremity Peripheral Nerve Quick Facts

MEDIAN NERVE:

- Innervates the abductor pollicis brevis (thumb abduction)

- Carpal tunnel is the <u>most common</u> compression neuropathy

ULNAR NERVE:

- Innervates the abductor digiti minimi (the little finger)

- Allows you to spread your fingers

RADIAL NERVE:

- Provides finger, wrist, elbow extension and a thumbs up!

AXILLARY NERVE:

- Injured with shoulder dislocations or humeral fractures

- Raises the shoulder above 15 degrees

Lower Extremity Peripheral Nerve Quick Facts

- Ilioinguinal nerve provides sensation to the medial thigh and genitals

- Lateral Femoral Cutaneous nerve provides sensation to the upper, lateral thigh — compression by obesity, pregnancy or heavy belts can cause *meralgia paresthetica*

- Femoral nerve provides for hip flexion and knee extension, as well as medial thigh and calf sensation. May be compressed by retroperitoneal hemorrhage.

- Obturator nerve provides for hip adduction, potentially injured during childbirth

Neuromuscular Quick Reference

Lower Extremity Quick Facts, continued

- Sciatic nerve innervates the adductor magnus and hamstrings (except the short head of the biceps femoris, which is the peroneal nerve!)

- Tibial nerve provides for foot *inversion* and plantar flexion

- Peroneal nerve provides for foot *eversion* and dorsiflexion (L5 nerve root dysfunction can also cause weakness of foot inversion)

Brachial Plexus

- Formed from the C5 – T1 nerve roots

- Lower plexus injuries may be associated with Horner Syndrome

- Lateral Cord (C5-C7) innervates elbow and wrist flexion, lateral forearm sensation

- Medial Cord (C7-T1) innervates thumb and finger flexors, intrinsic hand muscles and medial forearm sensation

- Posterior Cord (C5-C8) innervates extensor muscles and shoulder abduction, sensation of posterior hand and arm

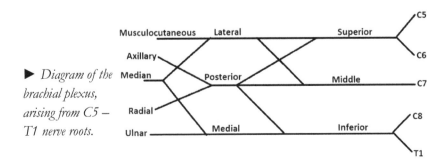

▶ *Diagram of the brachial plexus, arising from C5 – T1 nerve roots.*

Neuromuscular Quick Reference

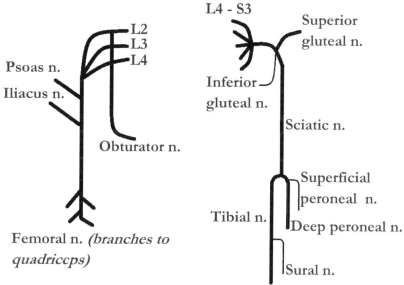

▲ *Diagram of the anterior lower limb nerves (left) and posterior lower limb nerves (right). L5 and S1 are the most commonly injured nerve roots in the lower extremities, usually from compression due to a herniated intervertebral disc. When the peroneal n. splits from the sciatic nerve, it wraps around the fibular head — a common site for compression*

Dermatomes of the hand and leg.

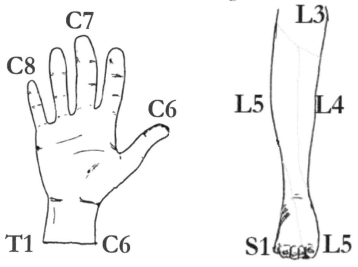

Neuromuscular Quick Reference Motor Exam Key

Movement	Root	Nerve	Muscle
Shoulder abduction	C5	Axillary	Deltoid
Elbow flexion	C5-6	Musculocutaneous	Biceps
	C6	Radial	Brachioradialis
Elbow extension	C7	Radial	Triceps
Wrist extension	C6	Radial	Extensor carpi
Finger extension	C7	Posterior interossei *(Radial n. branch)*	Extensor digitorum
Finger flexion	C8	Anterior interossei	Flexor pollicus
Finger abduction	T1	Ulnar	First doral interossei
		Median	Abductor pollicus brevis
Hip flexion	L1-2	Lumbar plexus	Iliopsoas
Hip adduction	L2-3	Obturator	Adductors
Hip extension	L5-S1	Sciatic	Gluteus maximus
Knee extension	L3-4	Femoral	Quadriceps
Knee flexion	S1	Sciatic	Hamstrings
Ankle dorsiflexion	L4	Deep peroneal	Tibialis anterior
Ankle eversion	L5-S1	Superficial peroneal	Peronei
Ankle plantarflexion	S1-2	Tibial	Gastrocnemius Soleus
Great toe extension	L5	Deep peroneal	Extensor halluces longus

Tip: When testing movements, try to support the limb so that you are isolating the movement and muscle of interest – this will help identify weakness that is due to a nerve or root cause.

Notes

Dementia

"Human sacrifice! Dogs and cats living together.
Mass hysteria!"
Ghostbusters

Cognitive Impairment and Dementia

Complaints about memory and thinking are common in the neurology clinic. Dementia affects about 10% of people over age 65 and almost half of people over age 85.

The most common cause of dementia is Alzheimer disease, which is a progressive neurodegenerative condition characterized by impaired memory. The second most common cause of dementia is cerebrovascular disease, usually from the accumulation of many small strokes and white matter disease. Alzheimer disease and vascular dementia are frequently occur together, and result in earlier and more severe memory loss.

There are many uncommon neurodegenerative conditions that cause dementia – several of them will be covered later in this chapter.

The Cognitive Exam

The history and patient interview are the core components of the memory evaluation. Many patients with dementia lose insight into their condition, making interviews with friends or family members invaluable. It is helpful to know the patient's baseline level of education and level of function – for example, what kind of work do they do? If they don't work, why not? What chores do they do at home, do they drive, do they have symptoms of depression, what medications do they use, and how have these changed in the past years?

The following are common complaints in memory loss and should be sought out:

- Memory loss affecting job skills
- Difficulty with familiar tasks
- Problems with language, word-finding difficulty
- Disorientation to time and place
- Poor or worsening judgment
- Difficulty with abstract thinking
- Misplacing items
- Changes in personality, mood or behavior
- Loss of initiative

There are several standardized tests available to screen for cognitive impairment. I recommend either the Folstein Mini-Mental Status Exam (MMSE), the Montreal Cognitive Assessment (MOCA), or the Memory Impairment Screen (MIS). Both the MOCA and the MIS are freely available on the internet. These are all great tests to screen for dementia. Note that results can be affected by acute illness or sedative medication, so they should ideally be administered when the patient is not acutely ill and is at their baseline functional status.

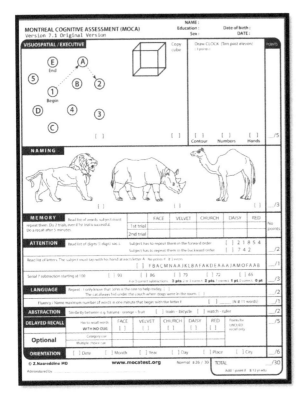

◀ *The Montreal Cognitive Assessment (MOCA) is a standardized memory test that is well suited to use in the clinic. It is freely available online. Different versions of the form exist, and it can be found in many languages. Available at www.MOCAtest.org.*

If significant cognitive impairment is found, patients can be referred for formal neuropsychological evaluation, which includes an extensive battery of cognitive testing under standardized conditions.

MIS: https://www.alz.org/documents_custom/mis.pdf

Case 1 – Dementia

A 55 year old right handed woman was a high functioning real estate agent without any significant prior medical history. Over the past 7 months she has developed progressive bizarre and erratic behavior. In the last three months she has become easily confused and has gotten lost in her own neighborhood. On exam she is tangential, has poor short term recall and has parkinsonism with rigidity and bradykinesia. She jumps and startles easily. There is no family history of dementia. A comprehensive metabolic panel is normal.

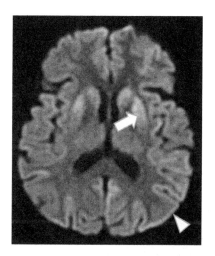

◄ *Diffusion weighted brain MRI, demonstrating abnormal bright signal along the cortex (arrowhead) and in the basal ganglia (arrow).*

This brain MRI, with diffuse cortical and basal ganglia diffusion restriction, is characteristic of Creutzfeldt-Jakob disease.

1. Does the presence of bright signal on DWI in the cortex and basal ganglia, as in this case, signify that the patient has had diffuse strokes?
A. Yes   B. No

2. True or False: Creutzfeldt-Jakob is caused by an infective agent?

3. The majority of cases of Creutzfeldt-Jakob in the United States are:___?
A. Acquired from contaminated surgical instruments
B. Acquired from transplantation with infected tissue
C. Acquired from eating infected beef
D. Hereditary
E. Sporadic

4. How is a <u>definitive</u> diagnosis of Creutzfeldt-Jakob made?
A. Autopsy   B. Brain MRI
C. CSF studies   D. EEG

Case 1 – Creutzfeldt-Jakob Disease

This patient has Creutzfeldt-Jakob Disease (CJD). It is caused by a brain protein with an abnormal structure, which spreads by inducing other brain proteins to acquire the same abnormal structure – thus making it an infectious protein. The vast majority of cases are sporadic, although there are cases acquired through contaminated surgical instruments and infected organ transplants (i.e. corneal transplant). When spread through contaminated beef it is known as Mad Cow Disease. It results in death over 6 – 12 months.

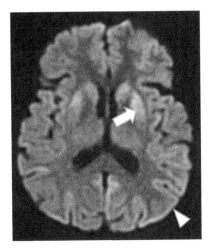

◄ *Diffuse cortical and basal ganglia diffusion restriction, as shown here, is almost pathognomonic of Creutzfeldt-Jakob disease in the right clinical context – however, the MRI may be negative early in the disease.*

1. Does the presence of bright signal on DWI in the cortex and basal ganglia, as in this case, signify that the patient has had diffuse strokes?
B. No – not all diffusion restriction is due to stroke, it is also seen in hyperdense tumors, acute demyelination, brain abscess and CJD

2. True or False - Creutzfeldt-Jakob is an infective condition.
A. True – it is a prion disease, due to an infectious protein

3. The majority of cases of Creutzfeldt-Jakob in the United States are:___?
E. Sporadic – 85% of cases are sporadic

4. How is a definitive diagnosis of Creutzfeldt-Jakob made?
A. Autopsy – brains should be evaluated at the National Prion Disease Pathology Surveillance Center at Case Western University

An elevated 14-3-3 protein in the spinal fluid is suggestive of CJD, but is not specific. The RT-QuIC is sensitive and very specific. EEG can be helpful as well, with a characteristic pattern of periodic discharges.

Case 2 – Dementia

A 68 year old man presents to clinic complaining of memory difficulty for the past year. He mostly notes difficulty with recalling the names of acquaintances and remembering recent conversations. He has a bachelor degree and worked as an accountant until three years ago, when he retired at age 65. He continues to pay his bills, maintain his home and drive around town. You perform a MOCA screening exam, and he scores 27/30, with points lost for delayed recall. He is otherwise well appearing without major medical issues. The rest of his neurologic exam is normal. He has been married for 45 years and has no high risk behavior.

1. Which of the following is not indicated in the routine work-up for dementia, according to American Academy of Neurology guidelines?
A. Serum B12
B. Rapid Plasma Reagin (RPR) for syphilis
C. Thyroid function studies
D. Depression screening
E. Liver function tests

2. Should brain imaging be obtained?
A. Yes – MRI brain with and without contrast
B. Yes – MRI brain without contrast
C. Yes – CT brain with contrast
D. No imaging indicated at this time

3. Should a lumbar puncture be routinely performed on all patients with suspected dementia?
A. Yes   B. No

4. Should an EEG be routinely performed on all patients with suspected dementia?
A. Yes   B. No

5. This patient's symptoms and exam most likely represent _____.
A. Normal aging
B. Mild cognitive impairment
C. Early dementia

Case 2 – Mild Cognitive Impairment

This patient does not have overt dementia, but does have memory impairment – specifically, he has a mild cognitive impairment (MCI). Approximately 10% of patients with MCI go on to develop dementia every year, although MCI is not a guarantee of developing dementia. In addition to the tests listed below, neuroimaging to rule out structural causes of confusion, such as strokes or tumors, is indicated.

Routine Evaluation for Dementia
- Complete blood count
- Glucose
- Serum electrolytes
- BUN/creatinine
- Depression screening
- Thyroid function tests
- Serum B12 levels
- Liver function tests

Note that testing for syphilis, HIV, heavy metal and lumbar puncture are only done if a clinical suspicion exists for specific illnesses.

1. Which of the following is not indicated in the routine work-up for dementia, according to American Academy of Neurology guidelines?
B. Rapid Plasma Reagin (RPR) for syphilis – HIV and RPR are not done routinely unless the history indicates exposure to these infections

2. Should brain imaging be obtained? **B. Yes – a non-enhanced MRI is preferred**

3. Should a lumbar puncture be routinely performed on all patients with dementia?
B. No – an LP is indicated if the condition is rapidly progressive, associated with early seizures/infectious symptoms, or otherwise atypical

4. Should an EEG be routinely performed on all patients with suspected dementia?
B. No – not unless you suspect subclinical seizures or witness overt seizures

5. This patients symptoms and exam most likely represent: **B. Mild cognitive impairment**

Tip: **Cholinesterase inhibitors** can improve behavior in mild to moderate dementia. **Vitamin E** may also be helpful at slowing the progression of symptoms.

See the AAN practice parameters for MCI:
http://tools.aan.com/professionals/practice/pdfs/dementia_guideline.pdf

Case 3 – Dementia

Five years later the same patient has returned to see you, with new memory complaints. His spouse states that he has significant short term memory problems, is frequently lost and is unable to take care of many of the basic household tasks he had previously performed. Last week he wandered away with the stove on and caused a small fire in the kitchen. He becomes easily angered and is less interactive. He is often awake through much of the night.

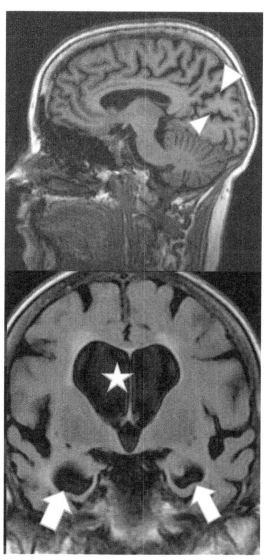

◄ *T1W sagittal MRI demonstrating significant cortical atrophy – look at the increased size of the sulci (arrowheads). The patient has gone on to develop Alzheimer dementia.*

◄ *T2W coronal FLAIR showing severe temporal lobe atrophy (arrows), which is due to loss of the hippocampus, a critical part of the brain for memory.*

Also note the size of the lateral ventricles (star). These appear dilated due to the severe volume loss, and is sometimes called 'hydrocephalus ex-vacuo.'

Case 3 – Alzheimer Dementia

Answers on the bottom of the page!

1. It is very unusual for Alzheimer Dementia to present before which age?
A. 50   B. 55   C. 60
D. 65   E. 70   F. 75

2. Which of the following is a common late complication of dementia?
A. Seizures
B. Non-ischemic cardiomyopathy
C. Sleep walking (somnambulism)
D. Loss of motor tone (hypotonia)
E. Enhanced sense of smell (hyperosmia)

This patient has typical symptoms of Alzheimer Dementia (AD), with memory loss and word finding difficulty. Atypical presentations of AD can include early visual and executive dysfunction. The typical age of onset is after 65 years, and incidence increases sharply with older age. Only 30% of cases of dementia before age 50 are due to AD.

Seizures are a common occurrence in advanced dementia, including AD – subclinical seizures should be considered when the demented patient has an abrupt change in level of consciousness.

Treatment is primarily supportive; cholinesterase inhibitors (i.e. donepezil) may help with mild to moderate symptoms. Memantine may improve function in moderate to severe cases. Many patients with dementia develop insomnia, which can worsen mood and fatigue. Advance care planning can be invaluable.

Restrictions on driving may vary from state to state, but must be addressed. Risk factors for car accidents include MMSE score of 24 or less, a caregiver's assessment of unsafe driving, self-restricted driving or a recent car accident.

Note that patients with trisomy 21 are very likely to develop AD pathology.

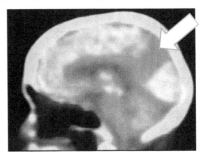

▶ *FDG-PET scans are nuclear medicine studies that measure brain metabolism. Decreased activity in the parietal (arrow) and temporal lobes (not shown) is consistent with a diagnosis of Alzheimer Dementia.*

Answers: 1: D, 2: A

Case 4 – Dementia

A 65 year old right handed man is in the intensive care unit for several weeks after being a passenger in a roll-over car accident. He suffered broken bones and internal bleeding, and his hospital course was complicated by ventilator associated pneumonia. He has gradually recovered and recently was taken off mechanical ventilation, but he is now found to be very easily confused and disoriented. On exam he doesn't know the date including the year, does not recognize that he is in a hospital, frequently becomes distracted and does not consistently follow commands. At times he is drowsy, and at other times he is more fluently conversant and seems to be better oriented.

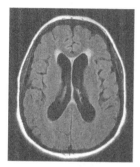

◄ *T2W FLAIR image shown here. The brain MRI showed no evidence of hemorrhage, no strokes, and only mild age related volume loss and white matter disease – a fairly unremarkable study.*

1. What is the most likely cause of the patient's fluctuating mental status?
A. Delirium   B. Traumatic brain injury
C. Dementia   D. Alcohol withdrawal

2. Which of the following is <u>not</u> a risk factor for developing delirium?
A. History of cognitive impairment
B. Sleep deprivation
C. Immobility
D. Vision or hearing impairment
E. Family history of delirium
F. Dehydration

3. Which of the following is <u>not</u> a potential precipitator of delirium?
A. Use of sedative drugs
B. Alcohol or drug withdrawal
C. Stroke
D. Fever or hypothermia
E. Metabolic derangements (i.e., electrolytes, glucose, etc.)
F. Surgery
G. Use of restrains
H. All of the above are precipitating factors in delirium

Case 4 – Delirium

This patient has delirium – an acute disorder of cognition and attention. It has been called "acute brain failure." Delirium is common, probably underdiagnosed, and can lead to significant morbidity and mortality. Delirium may lead to permanent cognitive decline, and has been shown to have long standing cognitive effects. The hallmark of delirium is acute (hours to days) onset of fluctuating confusion and alertness. It can be hyperactive with agitation and aggression, or more commonly, hypoactive. It is most commonly seen in the ICU and in post-operative patients.

Assessment of Delirium	
History	Baseline cognition, review all current medications, review alcohol and drug use, pain or discomfort
Vitals	Temperature, O2 sats, glucose levels
Exam	Focal deficits, dehydration, infection, abdominal pain, sensory impairment
Labs	Blood count, electrolytes, urinalysis, B12, TSH, ammonia, ABG, chest radiograph. Do lumbar puncture only if clinical suspicion for encephalitis exists.
Imaging	If any focal neurologic deficits exist or there is a history of head trauma
EEG	Assess for occult seizures, may help differentiate psychiatric disorders from delirium

Interventions to treat delirium include:

- Reducing psychoactive medications (sedatives, anticholinergics)
- Treat infections, metabolic disorders, dehydration, etc.
- Frequent reorientation, enlist family members to assist
- Mobilize patients as able, active range of motion
- Minimize restraints, including catheters, lines, drains, etc.
- Normalize sleep-wake cycle – discourage daytime napping
- Oral haloperidol 0.25-0.5 mg during the daytime may help if agitation is severe and threatens patient care. Quetiapine can also be effective.

1. What is the most likely cause of the patient's fluctuating mental status?
A. Delirium

2. Which of the following is <u>not</u> a risk factor for developing delirium?
E. Family history of delirium is not a risk factor, all of the others definitely are

3. Which of the following is <u>not</u> a possible precipitating factor for delirium?
H. All of the above are precipitating factors in delirium

Case 5 – Dementia

A 62 year old right handed woman has a history of migraine without aura and hypertension only. She is brought to the emergency department by her husband who states that she has suddenly become forgetful and confused – she cannot recall immediate events, and repeatedly asks the same questions. She introduces herself to you several times during the course of your exam, apparently not recalling have just met you. Her language, recall of long term events, and rest of the neurologic exam is completely normal however.

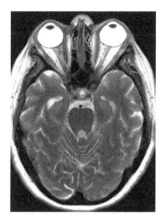

◀ *T2W brain MRI. There was no evidence of stroke, mass lesion or other abnormality.*

The patient also underwent an EEG which was unremarkable. Within 24 hours her symptoms had resolved and she had returned to baseline.

1. What is the most likely cause of the patient's memory complaint?
A. Stroke　　B. Psychogenic Amnesia
C. Delirium　　D. Transient Global Amnesia

2. Common triggers for this patient's condition includes all of the following except for which?
A. Severe vomiting　　　B. Cannabis use
C. Stressful life event　　D. Recent surgical procedure
E. Sexual intercourse

3. A stroke in which of the following brain areas is most likely to result in confusion and memory problems?
A. Hippocampus
B. Non-dominant frontal lobe
C. Cerebellar hemisphere
D. Internal capsule

Case 5 – Transient Global Amnesia

This patient has Transient Global Amnesia (TGA), which causes temporary impairment of short term (immediate) memory. Patient's cannot form new memories during an acute episode of TGA, but their memory for remote events is normal. The rest of the neurologic exam should be normal. Symptoms typically resolve within 12 – 24 hours.

A full neurological evaluation is essential to rule out other acute causes of impaired short term memory, such as stroke, seizure or encephalitis. Brain MRI and EEG are essential tests but are normal in TGA – a lumbar puncture should only be done if you have a clinical suspicion of infection, the symptoms don't resolve within 24 hours, or you find abnormalities on exam or imaging.

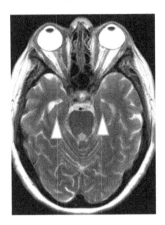

◀ *T2W brain MRI. The hippocampi are normal (arrowheads) in Transient Global Amnesia.*

The cause of TGA is unknown, but is thought to be due to hippocampal dysfunction – the hippocampi are essential structures for memory, especially the dominant (usually left brain) hippocampus. There is no specific treatment for TGA.

1. What is the most likely cause of the patient's memory complaint?
D. Transient Global Amnesia

2. Common triggers for this patient's condition includes all of the following except for which?
B. Cannabis use is the only one not implicated in TGA – Valsalva events such as intercourse, vomiting or exercise may be triggers

3. A stroke in which of the following brain areas is most likely to result in confusion and memory problems?
A. Hippocampus – usually due to a posterior cerebral artery stroke

Case 6 – Dementia

A 76 year old left handed man has an 18 month history of progressive urinary incontinence and gait instability. Over the past four months he has had increasing confusion, forgetfulness and has been less engaged with his family. On exam he has difficulty lifting his feet from the floor and very short steps. He turns slowly and his balance is poor. He had been previously healthy.

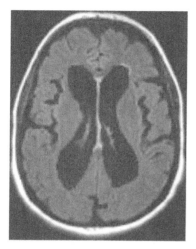

◄ *The patient's T2W FLAIR image is shown here. The radiologist notes enlarged ventricles, out of proportion to the amount of cerebral atrophy. There is no evidence of strokes, hemorrhages or asymmetric volume loss.*

1. The patient undergoes a lumbar puncture. Which of the following options would you most likely find in the CSF?
A. CSF pleiocytosis (increased white blood cells)
B. Positive HSV PCR
C. Severely elevated protein but normal white cell count
D. Normal CSF

2. The opening pressure on the LP was 20 mmHg. A large volume of CSF is removed during the lumbar puncture (> 30 ml), and the patient's urinary continence and gait temporarily improve. Which conditions is most likely?
A. Obstructive hydrocephalus
B. Alzheimer disease
C. Normal-pressure hydrocephalus
D. Chronic meningitis
E. Dementia pugilistic

3. Which of the following is the best long term treatment for this condition?
A. Thiamine supplementation
B. Ventriculoperitoneal shunting
C. Oral valacyclovir
D. Supportive care only

Case 6 – Normal-Pressure Hydrocephalus

This patient has normal-pressure hydrocephalus (NPH), which is an uncommon cause of dementia. The exact cause of NPH is unknown, but it is thought to be due to poor CSF reabsorption in the arachnoid granulations, and is a communicating form of hydrocephalus. Patients develop a triad of urinary incontinence (*wet*), gait disturbance (*wild*), and cognitive impairment (*wacky*). If patients show temporary improvement with a large volume lumbar puncture of lumbar drain, a ventriculoperitoneal shunt may be placed for long-term CSF diversion.

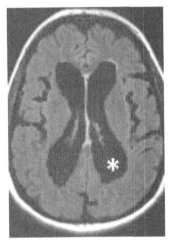

◀ *This T2W brain MRI shows dilated ventricles out of proportion to cerebral atrophy – suggestive of NPH.*

Normal CSF pressure is 10 - 20 mm Hg, and a typical CSF volume is 150 ml. About 500 ml of CSF are produced each day.

1. The patient undergoes a lumbar puncture. Which of the following options would you most likely find in the CSF?
D. Normal CSF – there is no inflammation or other CSF abnormality in NPH; abnormal CSF would strongly suggest an alternate diagnosis

2. The opening pressure on the LP was 20 mmHg. A large volume of CSF is removed during the lumbar puncture (> 30 ml), and the patient's urinary continence and gait temporarily improve. Which conditions is most likely?
C. Normal-Pressure Hydrocephalus typically has an opening pressure of 15-20 mm Hg, the upper range of normal; NPH may improve with early CSF removal

3. Which of the following is the best long term treatment for this condition? **B. Ventriculoperitoneal shunting – there are long term risks including infection and over-shunting (removal of too much CSF) which can lead to subdural hemorrhages**

Notes

Neuro Immunology

"What am I allergic to?"
- *"Pine nuts, and the full spectrum of human emotion."*
The Proposal

Multiple Sclerosis

The quintessential neuroimmunological condition is multiple sclerosis (MS), which is widely prevalent in the United States. Although MS is still incompletely understood, it is essentially an autoimmune condition in which an individual's white blood cells attack their own brain, optic nerves and spinal cord. Nerve axons, and their protective myelin coating (which function like the insulation on electric cables) are specifically targeted. The inflammatory lesions of MS are known as plaques.

The diagnosis of MS can be challenging, and many patients referred to MS clinics actually have a condition other than MS. However, the typical presentation involves multiple discreet neurologic symptoms spread out over time, also called 'dissemination in space and time,' meaning you have multiple lesions affecting different anatomical parts of the brain and spine, occurring at different times (see McDonald criteria, next page). For example, a typical pattern might include an episode of optic neuritis – inflammation of the optic nerve resulting in vision loss, followed by weakness and ataxia a year later.

The most common form of MS is the relapsing-remitting type (RRMS), in which patients have a clinical episode, for example of optic neuritis, and then experience a significant amount of recovery. Although patients improve between episodes, over time they accumulate significant disability.

Eventually many patients will develop a more progressive form of MS, in which they see little if any improvement between symptoms. Primary progressive MS starts this way from the beginning – these patients tend to be older at onset. In secondary progressive MS patients have lived with RRMS for years before their symptoms become progressive.

Recently there has been an explosion in immune modulating therapies for MS. However, a full review of these complex and highly specialized medications, used for long term disease control in MS, is outside of the purview of this book. We will discuss, rather, the acute treatment of new clinical events in MS, called MS flairs.

The primary treatment of new MS symptoms, for example new weakness or new optic neuritis, is high dose IV methylprednisolone. A common approach is to give three doses of once daily IV methylprednisolone 1,000 mg. For very disabling MS flairs some providers will treat with five days of IV steroids. Occasionally this will be followed by an 11 day oral prednisone taper, starting at 60 mg – however, this is not needed in most cases.

2017 McDonald Criteria for the Diagnoses of MS

Clinical Attacks	Number of MRI lesions	Additional data required for diagnosis
≥ 2	≥ 2	None
≥ 2	1	None – if attacks implicate different anatomical locations
≥ 2	1	Dissemination in space demonstrated by an additional clinical attack implicating another CNS site.
1	≥ 2	Dissemination in time demonstrated by an additional attack, by new MRI lesions or by oligoclonal bands
1	1	Dissemination in space shown by an additional clinical attack involving another anatomic location AND dissemination in time shown by MRI, oligoclonal bands or a new attack

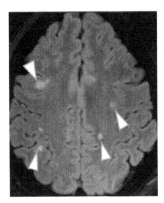

◀ *T2W FLAIR brain MRI in a patient with known multiple sclerosis, showing multiple lesions (arrowheads), many of which are adjacent to the cortex (juxtacortical).*

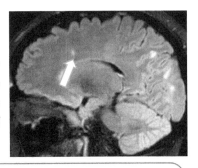

▶ *T2W sagittal FLAIR is a good way to see the periventricular lesions (arrow) in Multiple sclerosis. These lesions are often perpendicular the ventricle and are called Dawson's Fingers.*

Epidemiology: Peak age of onset for RRMS is 25 years, and women develop MS nearly twice as frequently as men. Men tend to develop MS at a slightly older age.

CASE 1 – Neuro Immunology

A 23 year old left handed woman with no prior medical history presents to the emergency department with progressive vision loss in her right eyes, which has gotten progressively worse in the last 24 hours. She has mild right eye pain when looking up and down or side to side. Her vision is fuzzy, "like someone smeared Vaseline over my eye." The funduscopic exam is unremarkable, with a sharp optic disc in both eyes.

1. You suspect that this patient's symptoms may be the first clinical attack of multiple sclerosis. What imaging study, if any, do you order?
A. Head CT with contrast
B. Non-enhanced brain MRI
C. Brain MRI with and without contrast
D. No imaging studies until a second event occurs

2. Would you perform a lumbar puncture?
A. Yes   B. No   C. Maybe

3. Which of the following is the best medical treatment for this patient at this time?
A. No treatment
B. IV methylprednisolone 1000 mg daily x 3 days
C. IV methylprednisolone 1000 mg daily x 5 days
D. Therapeutic plasma exchange x 5 days
E. IVIG

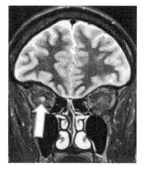

◀ *Her coronal T2W brain MRI demonstrated enhancement and increased signal in the right optic nerve (arrow). However, there were no other brain lesions on MRI.*

4. Given the brain MRI above, with no lesion other than the right optic nerve enhancement, what is the diagnosis?
A. Relapsing-remitting Multiple Sclerosis (RRMS)
B. Primary Progressive Multiple Sclerosis (PPMS)
C. Optic Neuritis
D. Optic Neuritis variant of Multiple Sclerosis

CASE 1 – Optic Neuritis

This is a classic presentation of optic neuritis – progressive loss of visual acuity and blurred vision developing over hours to days. Often there is mild pain with eye movements and a normal funduscopic exam. Optic neuritis refers specifically to inflammation of the optic nerve – although it is closely linked to Multiple Sclerosis and other neuroinflammatory conditions, it can occur in isolation.

Optic neuritis is the first sign of MS in about 15-20% of patients, and nearly half of patients with MS will have optic neuritis at some point.

Teaching Point: High dose IV steroids help patients with optic neuritis or MS attacks to recover more quickly, but may not impact the final degree of recovery.

1. You suspect that this patient's symptoms may be the first clinical attack of multiple sclerosis. What imaging study, if any, do you order?
C. Brain MRI with and without contrast – this is to assess for other brain lesions that would establish a diagnosis of MS

2. Would you perform a lumbar puncture?
C. Maybe – the LP can help establish a diagnosis of MS, but many providers consider it unnecessary in straightforward cases. I encourage a LP with oligoclonal bands and IgG index in all new cases.

3. Which of the following is the best medical treatment for this patient at this time?
B. IV methylprednisolone 1000 mg daily x 3 days

4. Given the brain MRI above, with no lesion other than the right optic nerve enhancement, what is the diagnosis?
C. Optic Neuritis – the MRI and history do not support dissemination in space or time… yet.

The Relative Afferent Pupillary Defect (rAPD)

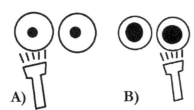

◀ *The rAPD is an important exam finding in optic neuritis. When you shine a flashlight into the unaffected eye (A), both eyes constrict. When you shin the light into the affected eye (B), both eyes dilate.*

CASE 2 – Neuro Immunology

The patient with optic neuritis from the previous case presents two years later with an episode of bilateral lower extremity weakness and urinary retention. On exam she has brisk reflexes in both legs, with an upgoing toe (extensor plantar response) on the right and several beats of clonus in the ankle on the right side. There is also an rAPD on the right, and vision is 20/40 in that eye.

▶ *Repeat MRI of the brain and spinal cord shows new lesions. In A) T2W FLAIR, there are multiple plaques (arrowhead). B) T2W sagittal spine MRI, demonstrates a new spinal cord lesion (arrow).*

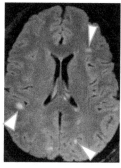
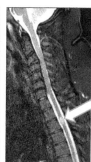

1. Given the patient's new symptoms and new imaging, which of the following is now the best diagnosis?
A. Relapsing-remitting Multiple Sclerosis (RRMS)
B. Primary Progressive Multiple Sclerosis (PPMS)
C. Optic Neuritis
D. Optic Neuritis variant of Multiple Sclerosis

2. Prior to receiving any treatment the patient states that she is pregnant. Are high dose IV steroids generally a safe treatment for new, severe MS attacks?
A. Yes, at any time
B. Yes, after the first trimester
C. Yes, after the second trimester
D. No, they are never safe

The patient is treated and has a significant recovery, despite residual lower extremity weakness and spasticity. Six months later she returns with complaints of recurrence of her prior leg weakness and urinary retention.

3. Which of the following is the next best step in the care of the patient given the recurrence of symptoms from a prior MS flare?
A. Treat with IV methylprednisolone 1000 mg daily x 3 days
B. Supportive therapy only
C. Obtain a urinalysis and work-up toxic/metabolic abnormalities

CASE 2 – Multiple Sclerosis

This patient has progressed from isolated optic neuritis to relapsing remitting multiple sclerosis, as her symptoms are now disseminated in space and time (see McDonald criteria). She should be referred to a neuroimmunologist for long term disease modifying treatment.

Spinal cord inflammation, called transverse myelitis, is common in multiple sclerosis and can cause significant disability – like the paraparesis (bilateral leg weakness) and urinary symptoms in this patient. Transverse myelitis due to multiple sclerosis usually only involves short segments of the spinal cord.

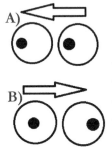

Internuclear Ophthalmoplegia (INO)

◀ *The INO is another common finding in multiple sclerosis. Due to disruption between the connection of the 6th and 3rd cranial nerve, patients may lose the ability to look to the side with both eyes (as in B). They are still able to look to the opposite side normally (as in panel A). This is due to a brainstem lesion, and can be from causes other than MS.*

1. Given the patient's new symptoms and new imaging, which of the following is now the best diagnosis?
A. Relapsing-remitting Multiple Sclerosis (RRMS)

2. Prior to receiving any treatment the patient states that she is pregnant. Are high dose IV steroids generally a safe treatment for new, severe MS attacks?
B. Yes, steroids are generally safe after the first trimester of pregnancy if patients have a disabling MS attack. In general, MS attacks are rare during pregnancy and increase temporarily after delivery.

3. Which of the following is the next best step in the care of the patient given the recurrence of symptoms from a prior MS flare?
C. Obtain a urinalysis and work-up toxic/metabolic abnormalities

Teaching point: The recurrence of prior symptoms can represent a 'pseudo-flare." That is, a metabolic or infectious insult causes a recurrence of symptoms from an old MS lesion. Steroids are not needed in that case – but the toxic/metabolic provoking factor should be identified and treated.

A few other Neuro Immunology tidbits...

Neuromyelitis Optica (NMO)

Another uncommon, but important disease to be aware of is NMO. This is an autoimmune condition that used to be considered a variant of MS, but is now recognized as an entirely separate disease.

NMO typically causes severe cases of optic neuritis and transverse myelitis, often with much less recovery than is seen in multiple sclerosis. Lesions in the brain are uncommon but do occur – more often than not within the brainstem itself.

Unlike MS, there is an antibody that can be sent to confirm the diagnosis of NMO. It is the anti-aquaporin 4 IgG antibody (also called the NMO antibody). This test should be sent from serum, not CSF.

▶ *T2W spine MRI, showing extensive transverse myelitis (arrows).*

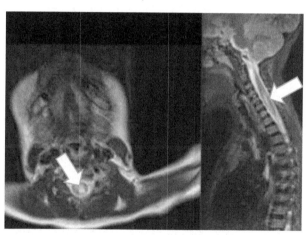

Transverse myelitis in NMO classically involves a larger segment of the spinal cord than do lesions in MS. A good rule of thumb is that a spinal lesion spanning three or more vertebral segments has a high likelihood of being due to NMO.

NMO does not respond to the typical treatments used in MS. Although high dose IV steroids can and should be used for acute symptoms in suspected or diagnosed NMO disorder, therapeutic plasma exchange may be necessary to remove excess circulating antibodies and minimize the progression of symptoms.

Common Medications used in the Treatment of MS

Glatiramer Acetate (Copaxone)
- Given as a subcutaneous injection, 3x weekly
- Pain or fat necrosis at the injection site can occur

Interferon (Avonex, Rebif)
- Given by weekly injection, or more frequently subcutaneously
- Can worsen depression, cause myalgias

Natalizumab (Tysabri)
- IV infusion and very effective MS treatment. In patients who carry the JC virus, there is a risk of developing PML (Progressive Multifocal Leukoencephalopathy), a lethal brain infection

Dimethyl fumarate (Tecfidera)
- Relatively new, very effective oral medication taken twice daily
- Well tolerated other than flushing and GI side effects

Ocrelizumab/Rituximab
- Monthly infusion, reduces B cell population

Fingolimod (Gilenya)
- Risk of heart block during medication initiation, avoid in patients with a history of cardiac disease

CSF Profile in Multiple Sclerosis

	WBC count Cells/mm^3	Protein Grams/dL	IgG Index	Oligoclonal Bands*
Normal	≤ 5 cells	< 45	Not elevated	Absent
MS	5 – 50 cells, mostly lymphocytes	Slightly elevated	Elevated in 75%	(+) in up to 95%
NMO	May be > 50 cells, can be neutrophils	Slightly elevated	Rarely elevated	(+) in 20%

* *Oligoclonal bands in the CSF means that there are antibodies present that are not found in the serum – it is a sensitive, but not specific, test for MS.*

Notes

Paul D. Johnson, MD

Behavioral Health

"Gentlemen, you can't fight in here! This is the war room!"
Dr. Strangelove or: How I learned to Stop Worrying
and Love the Bomb

The Mind-Brain Connection

There is a very close relationship between psychiatric health and brain health; neurologic and psychiatric illness are frequently interconnected and related in the same disease process. In fact, it's no accident that both neurologists and psychiatrists are certified by the same organization, the *American Board of Psychiatry and Neurology*. Although psychiatric illness can and does arise independently of neurologic illness, and vice-versa, if you consider that cognition, personality and the psyche are emergent phenomenon from the brain, then it makes sense that disruption of the central nervous system will result in disturbances of the psyche.

Many neurologic conditions are known to be associated with psychiatric illness. In general, patients with poorly controlled epilepsy, multiple sclerosis and degenerative brain diseases such as Alzheimer disease or Huntington disease are all found to have higher rates of depression, suicide and mental illness than comparable individuals without neurologic illness. This could, in part, be due to the fact that these conditions result in loss of brain volume over time, which in turn leads to emergence of psychiatric disease.

The relationship with neuroanatomy and psychiatric phenomenon can be illustrated by examining disease of the limbic system. The limbic system is comprised of several discreet structures within the brain that are associated with memory and behavior, including complex behaviors such as the sex drive, fear, anger and pleasure. The key limbic structures are connected through the *Papez Circuit*.

Papez Circuit

Internal capsule → Cingulate Cortex → cingulum → Hippocampus → fornix → Mammillary Bodies → Anterior Thalamic Nuclei → (Internal capsule)

Certain conditions have a predilection for the Papez circuit, especially the hippocampi, and are highly associated with behavior change. Herpes encephalitis typically affects limbic structures, and a key feature of early herpes encephalitis is irritability and bizarre behavior. Another example is anti-NMDA antibody encephalitis, an autoimmune condition leading to inflammation of limbic structures. Anti-NMDA antibody mediated encephalitis typically affects young women, and is frequently associated with ovarian teratomas, some of which can be microscopic. Neuropsychiatric effects are very common at the onset, with prominent behavioral changes and psychosis. Over time patients develop other findings such as orofacial dyskinesias and seizures.

Other important brain structures for behavioral health include the amygdala, located just anterior to the hippocampus. PTSD and anxiety are associated with increased activity within the amygdala. The nucleus accumbens is a dopaminergic structure within the basal ganglia, and is responsible for feeling pleasure.

Psychopharmacology

You will also note a significant overlap in medications used by both psychiatrists and neurologists, primarily the antipsychotics, antidepressants and antiepileptics.

Tricyclic antidepressants such as nortriptyline and amitriptyline are frequently used to treat neuropathic pain.

Antiepileptics such as valproic acid and lamotrigine are often used by psychiatrists as mood stabilizing agents, often at different doses than they would be prescribed at if used principally as antiepileptics. Levetiracetam, a commonly used antiepileptic, has primarily psychiatric side effects, including anger and irritability.

Antipsychotics, especially first generation (typical) medications, can cause secondary Parkinsonism and movement abnormalities such as tardive dyskinesia. Likewise, medications such as levodopa or the dopamine agonists can cause impulse control disorders or syndromes of excess movements, called dyskinesias.

Case 4 – Behavioral Health

A young, active adult notices progressive hand weakness, atrophy and dysarthria progressing over months. He is diagnosed with amyotrophic lateral sclerosis (ALS). One week later he has formed a local ALS support group, and immediately gets to work organizing a local 5K run/walk in support of ALS research. On race day, the participation was much lower than he had expected – that evening he kicked and yelled at his dog.

1. When the patient immediately responded to his ALS diagnosis by organizing a support group, he was demonstrating which type of emotional defense mechanism?
A. Transference
B. Sublimation
C. Usurpation
D. Intellectualization

2. When he beat his dog after the poor 5K race turnout, what type of emotional defense mechanism was he exhibiting?
A. Displacement
B. Projection
C. Reaction formation
D. Transference

3. Another young adult in the same town was diagnosed with ALS around the same time. He appeared calm, denied having strong emotions about his ALS diagnosis, and talked at length about the epidemiology of ALS and potential treatments. Which emotional defense mechanism was he exhibiting?
A. Sublimation
B. Self-observation
C. Intellectualization
D. Conversion

Case 4 – Defense Mechanisms

Defense mechanisms are unconscious methods of avoiding conscious conflict or anxiety – for example, rationalizing your way out of preparing for the neurology clerkship shelf exam. Defense mechanisms can result in both healthy and unhealthy behaviors, depending on how and when they are used. There is no exact consensus on the number of defense mechanisms, which are usually classified as mature, immature, neurotic or psychotic.

1. When the patient immediately responded to his ALS diagnosis by organizing a support group, he was demonstrating which type of emotional defense mechanism? **B. Sublimation – the transformation of unhelpful emotions into healthy actions**

2. When he beat his dog after the poor 5K race turnout, what type of emotional defense mechanism was he exhibiting? **A. Displacement – shifting aggressive impulses onto less threatening targets**

3. Another young adult in the same town was diagnosed with ALS around the same time. He appeared calm, denied having strong emotions about his ALS diagnosis, and talked at length about the epidemiology of ALS and potential treatments. Which emotional defense mechanism was he exhibiting? **C. Intellectualization – reasoning is used to block an unconscious conflict and associated emotional stress**

Common Defense Mechanisms

Dissociation	Dealing with stress by breaking down the usual integration of behavior, memory and perception
Reaction Formation	Formation of thoughts that are the opposite of the anxiety producing feelings
Projection	False attribution of one's own unacceptable feelings to another
Transference	The patient re-experiences the past in the setting of psychoanalysis
Counter Transference	The therapist's emotional entanglement with the patient
Conversion	Expressing a psychiatric conflict as a physical symptom, such as psychiatric blindness or non-epileptic spells

Case 5 – Behavioral Health

A 28 year old right handed woman has schizophrenia. Her symptoms have been unstable since her initial diagnosis five years prior. In clinic she complains of abnormal hair growth all over her body since starting a new antipsychotic medication, although the physical exam does not reveal any changes in hair growth. At the end of the visit, the patient states that she is 10 weeks pregnant.

1. Which additional symptom would support a diagnosis of paranoid type schizophrenia in this patient?
A. Posturing and immobility
B. Prominent hallucinations/delusions
C. Disorganized speech
D. Severely flattened affect

2. If this patient was acutely psychotic during labor, which medication would be safest to use?
A. Thioridazine
B. Haloperidol
C. Clozapine
D. Quetiapine

3. What are the chances of familial transmission of schizophrenia?
A. 0.1%
B. 1%
C. 5%
D. 10%

4. A 21 year old student with a history of well controlled anxiety suddenly becomes obsessed with the idea of an alien invasion, details of which are broadcast to him through the television. After two weeks the delusions and hallucinations resolve completely. Which of the following is the best diagnosis?
A. Schizophrenia
B. Schizoaffective disorder
C. Brief psychotic episode
D. Major depression with psychosis
E. Bipolar I disorder

> Neuroleptic Malignant syndrome (**NMS**) is a rare but potentially fatal side of effect of antipsychotic use. NMS is characterized by fever, muscle rigidity, altered mentation and autonomic instability. Treatment is supportive.

Case 5 – Schizophrenia

This patient has paranoid type schizophrenia. Schizophrenia is a syndrome involving chronic or recurrent psychosis, and patients usually have a constellation of positive and negative symptoms, mood or anxiety disorders, disorganized or catatonic behavior, and cognitive impairment.

1. Which additional symptom would support a diagnosis of paranoid type schizophrenia in this patient?
B. Prominent hallucinations/delusions are most suggestive of paranoid schizophrenia

2. If this patient was acutely psychotic during labor, which medication would be safest to use?
B. Haloperidol – high-potency typical antipsychotics may be appropriate for use in active psychosis during delivery, especially if the psychotic behavior poses a risk to the mother and child

3. What are the chances of familial transmission of schizophrenia?
D. 10%

4. A 21 year old student with a history of well controlled anxiety suddenly becomes obsessed with the idea of an alien invasion, details of which are broadcast to him through the television. After two weeks the delusions and hallucinations resolve completely. Which of the following is the best diagnosis? **C. Brief psychotic episode – a brief psychotic disorder lasts between 1 day and 1 months; whereas the symptoms of schizophrenia must be present for at least 6 months to make a diagnosis**

Clinical Manifestation of Schizophrenia	
Positive	Hallucinations, delusions or disorganized thought (tangential, circumstantial speech, word salad, etc.)
Negative	Deficit symptoms such as flat affect, apathy, lack of energy – these can be very difficult to treat
Mood	Depression and anxiety are highly prevalent in schizophrenia
Cognition	Neuropsychological testing usually 1-2 standard deviations below normal
Catatonia	Either extreme negativism (mute, unmoving) or excessive, purposeless motor activity – treat with benzodiazepines

Behavioral Health – Withdrawal Syndromes Quiz

Match the substance listed below (A-F) to the correct description of that substance's withdrawal symptoms. Each one has a single best answer and none are repeated.

A. Alprazolam
B. Heroin
C. Ethanol
D. Paroxetine
E. MDMA (ecstasy)
F. Cocaine

1. The most dangerous withdrawal syndrome with up to 75% having a generalized seizure by the third day of withdrawal

2. May cause depression and sleeplessness within days of stopping

3. Intense dysphoria, fatigue, hypersomnia and increased appetite

4. Very short half-life can lead to symptoms of panic

5. Bruxism, hyperthermia, diaphoresis, tachycardia

6. Piloerection, diaphoresis, yawning, cramps, tearing

Behavioral Health - Withdrawal Syndromes Quiz

Match the substance listed below (A-F) to the correct description of that substance's withdrawal symptoms (1-6). Each one has a single best answer and none are repeated.

   A. Alprazolam
   B. Heroin
   C. Ethanol
   D. Paroxetine
   E. MDMA (ecstasy)
   F. Cocaine

1. The most dangerous withdrawal syndrome with up to 75% having a generalized seizure by the third day of withdrawal
   C. Ethanol

2. May cause depression and sleeplessness within days of stopping
   D. Paroxetine (shortest half-life SSRI)

3. Intense dysphoria, fatigue, hypersomnia and increased appetite
   F. Cocaine (may also have suicidal ideation)

4. Very short half-life can lead to symptoms of panic
   A. Alprazolam (very short half-life benzodiazepine)

5. Bruxism, hyperthermia, diaphoresis, tachycardia
   E. MDMA

6. Piloerection, diaphoresis, yawning, cramps, tearing
   B. Heroin

Notes

Paul D. Johnson, MD

Neurosurgery

"It's naht a toomah!"
Kindergarten Cop

CASE 1 – Neurosurgery

A 35 year old woman with polycystic kidney disease, hypertension and tobacco use has a sudden onset, severe headache which occurred abruptly while exercising. Within minutes she has a brief, generalized tonic-clonic seizure and loss of consciousness. When EMS arrives on the scene her Glasgow Coma Scale (GCS) is 3, and she is intubated for airway protection and brought to the ER. Her blood pressure is very elevated at 190/120 mmHg.

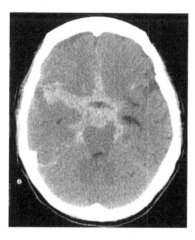

◄ *Non-enhanced head CT, obtained in the Emergency Department.*

1. Which of the following best describes the findings from the above CT?
A. There is diffuse cerebral edema
B. There is a tumor
C. There is diffuse subarachnoid hemorrhage
D. There is a large subdural hemorrhage
E. This is a normal head CT

2. Which of the following is the best immediate treatment for this patient?
A. Administer rectal aspirin
B. Allow permissive hypertension
C. Open craniotomy and hemorrhage evacuation
D. Blood pressure lowering below 160 mmHg
E. Placement of a lumbar spine drain

3. For patients who present with a thunderclap headache (maximal onset within 60 seconds), but have a negative head CT within 6 hours of presentation, what is the next best diagnostic step?
A. Lumbar Puncture   B. Repeat CT in 24 hours   C. Watchful waiting

CASE 1 – Subarachnoid hemorrhage

This patient has a large subarachnoid hemorrhage (SAH). Although SAH can occur from trauma, or rupture of vascular malformations such as arteriovenous malformations, by far the most common cause is the rupture of a sacular (or 'berry') aneurysm.

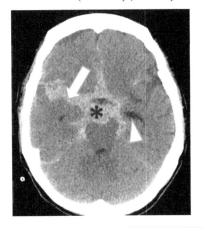

◀ *This non-enhanced head CT demonstrates significant subarachnoid hemorrhage within the basal cisterns (*) between the frontal and temporal lobes (arrow). There is early hydrocephalus with expansion of the temporal pole of the left ventricle (arrowhead).*

Ottawa Subarachnoid Hemorrhage rule to identify patients ≥ 15 years old who need an LP to rule out SAH. Does not apply to people with prior aneurysm or chronic headache. Headache onset must be maximal within the first hour.
- Age ≥ 40
- Neck pain or stiffness
- Witnessed loss of consciousness
- Onset during exertion
- Thunderclap headache (maximal within 60 seconds)
- Limited neck flexion on exam

1. Which of the following best describes the findings from the above CT?
C. There is diffuse subarachnoid hemorrhage

2. Which of the following is the best immediate treatment for this patient?
D. Blood pressure lowering below 160 mmHg

3. For patients who present with a thunderclap headache (maximal onset within 60 seconds), but have a negative head CT within 6 hours of presentation, what is the next best diagnostic step? **A. Lumbar Puncture to rule out SAH – Head CT can miss cases after 6 hours**

Tip: There are several classification schemes to measure the severity of a SAH. The most commonly used are the Hunt and Hess (H&H) and World Federation of Neurosurgeons (WFNS) scales.

CASE 1 – Subarachnoid hemorrhage

After being stabilized in the emergency department, lowering the blood pressure below 160 mmHg and starting an antiepileptic medication, the patient undergoes a CT angiogram to look for the source of the subarachnoid bleeding.

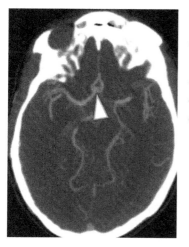

◄ *This CT angiogram identified an anterior communicating sacular aneurysm (arrowhead). Almost 1/3 of all aneurysms will be found at the anterior communicating artery – at the origin of the two anterior cerebral arteries.*

1. In general, for patients with subarachnoid or intraparenchymal hemorrhage, which of the following is the most appropriate imaging study?
A. CT angiogram of the head only
B. CT angiogram of the head and neck only
C. CT venogram
D. CT angiogram of the head, neck and CT venogram

▶ *A conventional catheter angiogram was obtained next, demonstrating the anterior communicating artery sacular aneurysm (arrowhead). Note the internal carotid artery (*) and anterior cerebral arteries (arrow).*

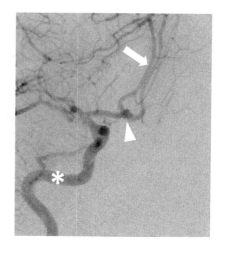

Answer: 1) A. CT angiogram of the head only – for brain hemorrhages, angiography of the head alone is usually sufficient.

CASE 1 – Subarachnoid hemorrhage

This patient has a ruptured anterior communicating artery aneurysm. Unsecured (i.e. untreated) aneurysms are very dangerous and have a high rate of recurrent bleeding, with a high mortality rate. The priority now is treatment. There are two options – clipping is an open brain surgery where a metal clip is placed at the base of the aneurysm, effectively sealing it off. This is a very effective, durable treatment but requires a craniotomy and some aneurysms are easier to access than others.

Second is coiling, in which a long, thin metal filament is placed into the aneurysm using an endovascular technique. This effectively fills the aneurysm, filling it with thrombus, and preventing aneurysm growth and rupture. Not all aneurysms are amenable to coiling, depending on their shape, but this minimally invasive technique is becoming widely used.

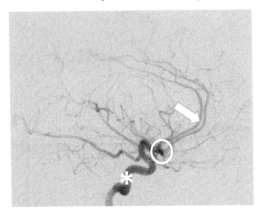

◄ *Another view from the same patient's conventional catheter angiogram, again showing the aneurysm (in circle), internal carotid artery (*) and anterior cerebral arteries.*

Because this patient's aneurysm was easily accessible by surgery, she underwent surgical clipping, with good results.

► *A) Digital reconstruction showing clipped aneurysm (arrow). B) Post-operative head CT with pneumocephalus (*) and mesh at craniotomy site (arrowhead).*

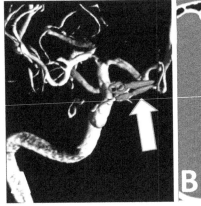
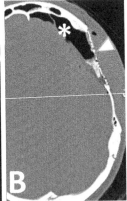

CASE 1 – Subarachnoid hemorrhage

Although this patient's aneurysm was effectively treated by surgical clipping, she remained in the neurosurgical intensive care unit for another two weeks for evaluation and treatment of complications of subarachnoid hemorrhage. During that time she underwent daily Transcranial Doppler (TCD) evaluation to assess for vasospasm.

1. Which of the following is the next best diagnostic step for patients who have a subarachnoid hemorrhage, but in whom no aneurysm is found on catheter angiogram (answers at bottom)?
A. No further diagnostic studies are required
B. Repeat a conventional catheter angiogram within one week
C. Repeat a conventional catheter angiogram in one year
D. Repeat a conventional catheter angiogram if patient has new SAH

Potential Complications from Subarachnoid Hemorrhage

Vasospasm
- May be severe enough to cause delayed ischemic strokes
- Evaluate with daily Transcranial Doppler ultrasound
- May require endovascular vasodilators or angioplasty

Hydrocephalus
- Most common complication after SAH
- May require a ventriculostomy and CSF shunting

Epilepsy
- Not all patients with acute seizures (within 1 week of SAH) will develop epilepsy – although some will
- Use antiepileptics for at least 6 months following an acute seizure to prevent recurrence

Hyponatremia
- Frequent, likely due to SIADH in euvolemic patients
- Rarely due to cerebral salt wasting, causes hypovolemia

Answer: 1) B. Repeat a conventional catheter angiogram within one week

CASE 1 – Subarachnoid hemorrhage

Let's provide an example of a patient who also presented with SAH, but who was treated with endovascular coiling. This patient was 45 years old and had an acute onset of headache and decreased level of consciousness, and was found to have a ruptured basilar tip aneurysm.

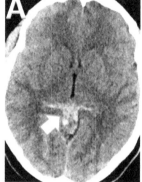

◄ A) Non-enhanced head CT showing subarachnoid hemorrhage (white arrow).

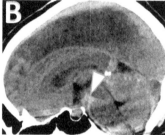

◄ B) Sagittal view of the same non-enhanced head CT with SAH (arrowhead).

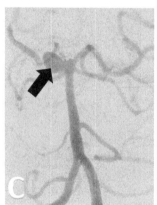

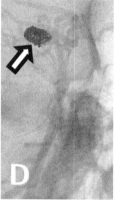

◄ C) Catheter angiogram showing the basilar tip sacular aneurysm (solid arrow). D) Catheter angiogram showing the successfully coiled aneurysm (outlined arrow), a minimally invasive endovascular procedure.

This patient went on to do very well after aneurysm coiling.

Teaching point: Small aneurysms are sometimes found incidentally, and can be monitored with serial imaging. Any aneurysm should be evaluated by a neuro-interventionalist.

CASE 2 – Neurosurgery

A 19 year old right handed woman is seen for chronic, severe headaches. She has no other medical problems, and no family history of headaches. When she coughs, picks up heavy bags, or uses the bathroom the headaches become worse. He also complains of significant neck pain. Exam is notable only for slightly brisk lower extremity reflexes and slight ataxia on finger to nose exam. An outpatient brain MRI was ordered.

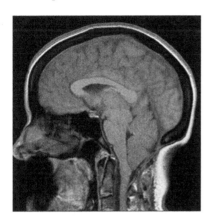

◀ *T1W sagittal brain MRI.*

1. Which of the following best describes the findings from the above MRI?
A. Normal brain MRI
B. There is hydrocephalus
C. The cerebellar tonsils are abnormally low
D. There is malformation of the corpus callosum
E. There is a tumor in the brainstem

2. For this patient, who is actively experiencing symptoms, which of the following is the next best step in treatment?
A. Referral to neurosurgery for possible decompression
B. Schedule acetaminophen until symptoms improve
C. Arrange for radiation and chemotherapy
D. Place a ventriculostomy to relieve the hydrocephalus

CASE 2 – Chiari 1 Malformation

This young adult has a type 1 Chiari malformation, which means that the cerebellum is unusually low lying, and actually herniating through the foramen magnum (the base of the skull). Essentially, there isn't enough room in the skull for the cerebellum and it's being squeezed out – potentially putting pressure on the spinal cord.

Diagnosing a Chiari malformation can be more difficult than it sounds – a small amount of herniation (up to 3mm) can be normal, and treating a Chiari when asymptomatic is controversial. Typically, a patient should have at least **5mm** of cerebellar herniation and symptoms (headache, neck pain, spinal cord dysfunction or cranial neuropathy) to warrant treatment.

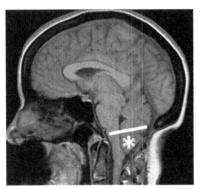

◄ *T1W sagittal brain MRI showing the low lying cerebellum (*) below the foramen magnum (white line). Notice the pressure this is having on the spinal cord. Compare this to the normal cervical spine on page 20.*

Treatment for this condition, when symptomatic, is surgical decompression, including sub-occipital craniotomy, giving the cerebellum more room. They may also perform a laminectomy of the fist cervical vertebrae.

1. Which of the following best describes the findings from the above MRI?
C. The cerebellar tonsils are abnormally low

2. For this patient, who is actively experiencing symptoms, which of the following is the next best step in treatment?
A. Referral to neurosurgery for possible decompression

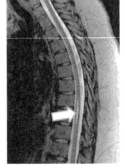

► *Patients with Chiari malformation may also develop a syrinx (arrow), a fluid filled space in the middle of the spinal cord. This severe complication of Chiari can cause myelopathy (spinal cord dysfunction), especially pain.*

CASE 3 – Neurosurgery

A 25 year old, previously healthy, right handed man was an unrestrained passenger in a high speed motor vehicle accident. On the scene of the accident he was comatose and unresponsive, with reflexive posturing on exam. He was intubated for airway protection, but did not require heavy doses of sedatives.

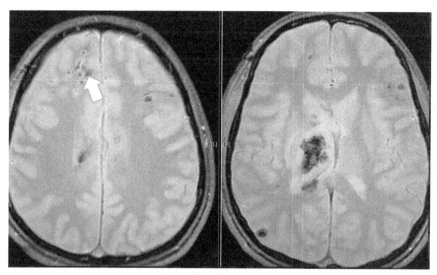

▲ *This is a GRE brain MRI (one of the blood & calcium sensitive T2* sequences), showing diffuse hemorrhage, consistent with a high velocity shear injury.*

1. Which of the following best describes the finding indicated by the white arrow on the MRI shown above?
A. Acute ischemic stroke
B. Necrotic brain tissue
C. Microscopic hemorrhage
D. Parenchymal hematoma

2. Some patients with severe traumatic brain injuries have no clear injury, including no bleeding, on head CT. In that case, is there any utility to obtaining a brain MRI?
A. No, any bleeding would be obvious on CT
B. No, MRI has no prognostic value
C. Yes, some injuries such as microscopic hemorrhage from sheer injury are only seen on MRI

CASE 3 – Diffuse Axonal Injury

This young man has a severe traumatic brain injury – the high velocity impact results in sheering – or tearing – of the axons. Remember that axons are the long, thin parts of neurons that connect brain cells and make up the white matter. The traumatic tearing of axons is called a diffuse axonal injury (DAI). Although the head CT may look relatively normal, patients with DAI may have significant amounts of microhemorrhage on T2* weighted blood sensitive MRI (such as SWI, or the GRE shown here). These patients are often comatose. Given the diffuse, severe injury, prognosis is often poor.

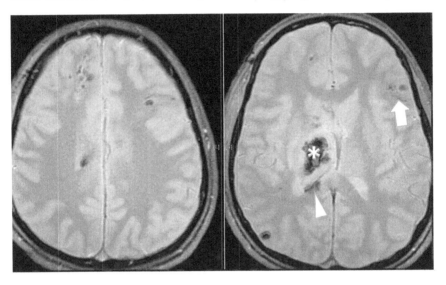

▲ *This is GRE brain MRI shows diffuse microhemorrhages, including within the corpus callosum (arrowhead), basal ganglia (*), and the junction of the grey and white matter (arrow) - common locations for shear injury.*

1. Which of the following best describes the finding indicated by the white arrow on the MRI shown above?
C. Microscopic hemorrhage

2. Some patients with severe traumatic brain injuries have no clear injury, including no bleeding, on head CT. In that case, is there any utility to obtaining a brain MRI?
C. Yes, some injuries such as microscopic hemorrhage from sheer injury are only seen on MRI – microhemorrhages may not be visible on CT scan. In severe hemorrhage, there may also be delayed hemorrhage.

CASE 4 – Neurosurgery

A 37 year old, right handed woman has recently been diagnosed with a left temporal lobe brain tumor. Her primary symptoms correlate with the tumor location – mild difficulty with language and right face and arm weakness. She comes to the clinic today to discuss treatment and options, and has not yet started any therapy.

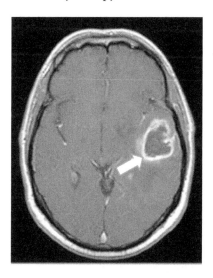

◄ *T1W post-contrast brain MRI showing a ring enhancing lesion (arrow) in the left temporal lobe.*

1. The patient has not had a seizure, but asks about starting empiric antiepileptic medication. Based on American Academy of Neurology guidelines, what should you tell her?
A. Empiric antiepileptic treatment is not indicated
B. Empiric antiepileptic medication should be started now
C. Empiric antiepileptic is indicated for certain tumor types only

2. You suspect the patient has a Glioblastoma (GBM). Which of the following is the standard of care for GBM tumors?
A. Radiation therapy alone
B. Radiation plus temozolamide chemotherapy
C. Surgical resection, radiation and temozolamide chemotherapy

3. Many patients with brain tumors receive treatment with dexamethasone. Which of the following best describes the benefit of oral steroids for GBM?
A. Actively kills tumor cells
B. Limits tumor growth
C. Decreased tumor associated edema

CASE 4 – Glioblastoma Multiforme

This patient does have a GBM, a very aggressive brain tumor with an average survival of around 12 – 18 months. Almost half of all primary brain tumors in adults will be GBM. However, adult brain tumors are more likely to be metastatic from another location – usually breast or lung cancer.

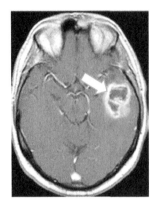

◀ *T1W post-contrast brain MRI with left temporal Glioblastoma Multiforme (GBM, arrow). These high grade brain tumors often show central necrosis and a ring of enhancement. They often cross midline and grow along white matter tracts, such as the corpus callosum. Tumor cells generally spread far beyond what is seen on the MRI.*

1. The patient has not had a seizure, but asks about starting empiric antiepileptic medication. Based on American Academy of Neurology guidelines, what should you tell her?
A. Empiric antiepileptic treatment is not indicated – although the AAN does not recommend starting antiepileptic medication for patients with brain tumors who have never had seizures, clinical practice does vary and some providers do use AEDs in these patients.

2. You suspect the patient has a Glioblastoma (GBM). Which of the following is the standard of care for GBM tumors?
C. Surgical resection, radiation and temozolamide chemotherapy

3. Many patients with brain tumors receive treatment with dexamethasone. Which of the following best describes the benefit of oral steroids for GBM?
C. Decreased tumor associated edema. Steroids should be used very cautiously in undiagnosed brain masses, as they will confound the diagnosis of primary CNS lymphoma.

Teaching point: Brain tumors, especially GBM, promotes thrombosis and these patients have a high rate of DVT and risk for pulmonary embolism.

Also recall, that brain tumors such as GBM are a contraindication for treatment with IV tPA.

Ring Enhancing Lesions

Many brain tumors show enhancement on post-contrast brain MRI. They can be referred to, non-specifically, as ring enhancing lesions. Let's take a look at what's included in the differential diagnosis for ring enhancing lesions:

Metastasis: the most common tumors in the brain are actually metastasis from other locations, especially lung and breast.

Abscess: infectious abscesses have ring enhancement, lots of surrounding vasogenic edema, *and* the necrotic core will diffusion restrict.

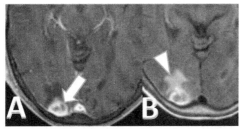

◄ *A) T1W post-contrast MRI showing a ring enhancing abscess (arrow).*

B) T2W FLAIR showing vasogenic edema (arrowhead).

Glioblastoma: the enhancing ring represents new blood vessel growth in the periphery of the tumor.

Infarct: strokes actually have *subacute* peripheral enhancement due to recruitment of new blood vessels around the stroke bed.

Contusion: similar to the subacute enhancement in infarcts, it reflects the healing process.

Demyelination: enhancement in MS lesions reflects their acute nature. Often seen found as an 'open ring' or C shaped enhancement.

Radiation Necrosis: less common nowadays.

Use the mnemonic is **MAGIC DR** to help you remember the list.

Brain Tumors

There are many different types of brain tumors, but here are some of the common ones that you should be familiar with.

Meningiomas are slowly enlarging tumors that grow out of the meninges. Because of the slow rate of growth, they can be quite large at the time of diagnosis. Although 'benign' because they don't cause distant metastasis, they can cause significant morbidity and mortality if in a difficult to remove location, like the base of the brain. While mostly sporadic, they can also develop after brain radiation.

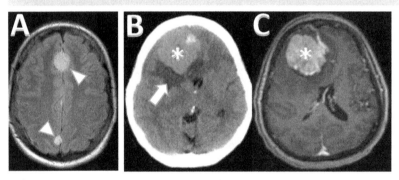

▲ *A) T2W brain MRI showing two midline meningiomas (arrowheads). B) Non-enhanced head CT showing hyperdense mass (*) with surrounding vasogenic edema (arrow), and C) T1W post-contrast MRI of the same patient showing the clearly defined meningioma (*). Note the mass effect on the ventricles.*

Schwannomas are tumors that arise from the auditory nerve, and are the most common tumor in the posterior fossa in adults. They may cause hearing loss and tinnitus, and are associated with neurofibromatosis type 2, which is due to a mutation on chromosome 22.

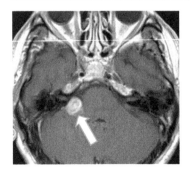

◄ *T1W post-contrast MRI, demonstrating an enhancing mass in the right pontocerebellar angle (arrow). This is a vestibular schwannoma, and the patient had progressive right sided sensorineural hearing loss and tinnitus.*

Brain Tumors

Ependymomas are glial tumors that arise from the ependymal cells lining the ventricles. They tend to develop in the brain, and especially within the posterior fossa, in children. The typical presentation is related to increased intracranial pressure, with nausea, headache, papilledema, ataxia and vertigo. In adults Ependymomas typically arise within the spinal cord. Treatment is typically surgical resection followed by radiation. Pseudorosettes are present on histopathology.

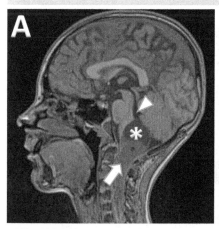
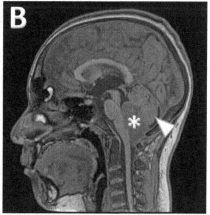

▲ *A) T1W brain MRI demonstrating a posterior fossa ependymoma with spread into the upper spinal cord (arrow). Not the cystic space (*) and fourth ventricle (arrowhead). B) T1W MRI of another patient with ependymoma. Note the tumor clearly within the fourth ventricle (*) and mass effect on the cerebellum (arrowhead).*

Rathke Cleft Cysts are benign cysts found near the pituitary, arising from the embryologic remnants of Rathke's pouch in the pituitary. Most are incidental, although if large they can cause headache or pituitary dysfunction. The main differential is a **Craniopharyngioma**, a slow growing mixed-density tumor also arising from Rathke's pouch.

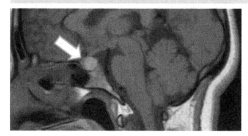

◄ *T1W MRI demonstrating a Rathke cleft cyst (arrow). The appearance is variable, but they do not enhance. Craniopharyngiomas are more likely to be calcified and to enhance.*

Brain Tumors

Medulloblastoma is the <u>most common</u> malignant brain tumor in children. It typically presents as a midline mass within the fourth ventricle, arising from the cerebellar vermis 75% of the time. Compression of the fourth ventricle results in obstructive hydrocephalus and intracranial pressure. Medulloblastoma does not tend to 'squeeze through' the foramen of Lushka, unlike an ependymoma. On histology these tend to be 'small round blue cell' tumors.

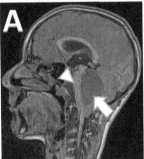 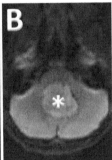

◄ A) T1W brain MRI, with Medulloblastoma filling the fourth ventricle (arrow) and compressing the brainstem (arrowhead). B) Axial DWI showing Medulloblastoma within the fourth ventricle (*), with diffusion restriction.

Oligodendogliomas frequently arise in middle aged adults (40-50 years of age), and often involve the cortex or subcortical white matter. Due to the cortical involvement there is a high risk of seizures. These tumors tend to calcify, and have high rates of hemorrhage. Histologically the cells look like 'fried eggs' and the presence of fine capillaries (hence the bleeding risk) gives a 'chicken wire' appearance.

► A) T2W brain MRI, demonstrating an expansile hyperintense, subcortical lesion (arrowhead). B) T2* MRI shows an area of microhemorrhage (arrow), consistent with the high bleeding risk of oligodendroglioma.

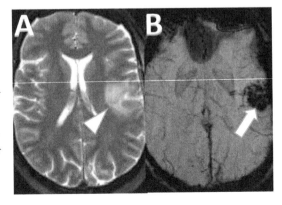

CASE 5 – Neurosurgery

A 68 year old woman is well other than being a smoker and having been involved in a motor vehicle accident two years ago. For the past six months she has complained of bilateral arm pain and numbness in her fingers, as well as clumsiness with buttons. Over the past month she has had several falls, which she had attributed to leg stiffness. Now, she is brought to the emergency department after a recent fall resulted in sudden, significant weakness in all extremities. You order a bladder scan which shows 500cc's of urine.

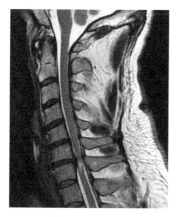

◀ *T2W sagittal cervical spine MRI. The MRI report states that there is multilevel disc disease, narrowing of the foramina, and spinal cord signal change.*

1. Which of the following is the cause of the patient's sudden symptoms?
A. Spinal cord infection
B. Spinal cord infarction
C. Spinal cord compression
D. Autoimmune spinal cord inflammation

2. Which of the following options is the best acute treatment for this patient?
A. High dose steroids
B. Antibiotics and supportive care
C. Aspirin and blood pressure control
D. Surgical decompression

3. Which of the following are you most likely to find on this patients exam?
A. Brisk reflexes in the legs, extensor toes
B. Flaccid weakness, extensor toes
C. Reduced leg reflexes, down going toes
D. Brisk leg reflexes, extensor toes

Case 5 – Cervical Spondylotic Myelopathy

This patient has spondylosis, or degenerative spine disease, leading to acute on chronic cervical spinal cord compression and myelopathy. Cervical spondylotic disease is the most common cause of spinal cord dysfunction in the those over age 55. Progressive dysfunction with gait is one of the most common presenting symptoms in cervical spondylotic myelopathy, as are leg stiffness, hand clumsiness, and paresthesias in the fingers and legs. Conservative management includes spine immobilization and NSAIDs.

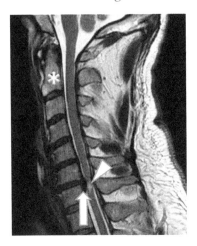

◄ *T2W sagittal cervical spine MRI showing a posteriorly herniated cervical disc (arrow) at C6 – C7. The presence of focal T2W cord signal change (arrowhead) indicates edema and cord injury. Note that the first visible vertebra (*) is C2, as C1 is not well visualized on axial MRI. A spine CT will not show cervical cord pathology, and non-contrast MRI is preferred.*

1. Which of the following is the cause of the patient's sudden symptoms?
C. Spinal cord compression

2. Which of the following options is the best acute treatment for this patient?
D. Surgical decompression is best for severe symptoms, especially given the presence of cord signal change. The goal of surgery is to stop progression and preserve remaining neurologic function.

3. Which of the following are you most likely to find on this patients exam?
A. Brisk reflexes in the legs, extensor toes – you would expect upper motor neuron signs.

Teaching Point: Even if a patient's symptoms seem to localize to the lower spinal cord, <u>the lesion may still be in the cervical cord</u> – and the C-spine should be imaged!

Further reading: Tavee JO, Levin KH. Myelopathy Due to Degenerative and Structural Spine Diseases. Continuum (Minneap Minn) 2015;21(1): 52-66.

Case 6 – Neurosurgery

A 55 year old right handed man was admitted for a hemorrhagic stroke, with intraventricular extension of the hematoma. Although initially alert and interactive, over the next 24 hours, he became lethargic and difficult to arouse. A repeat non-enhanced head CT showed increased size of his lateral ventricles, and blood within the posterior fossa. A neurosurgeon placed an external ventricular drain (EVD) at bedside.

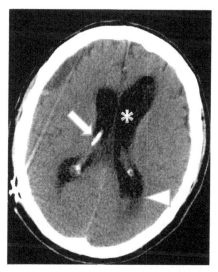

◄ *Non-enhanced head CT, with enlarged lateral ventricles, consistent with acute hydrocephalus (*). You can see the tip of the EVD (arrow) in the right lateral ventricle. Note the fluid level in the bottom of the left ventricle (arrowhead), representing subacute blood pooling there.*

1. What is the most likely reason why this patient developed hydrocephalus?
A. Hemorrhage induced inflammation
B. Acute increase in CSF production
C. Obstruction of CSF drainage due to blood in the ventricles

2. Which of the following symptoms are <u>commonly</u> caused by acute obstructive hydrocephalus?
A. Decreased level of consciousness
B. Acute seizures
C. Thunderclap headache

3. EVD's are generally placed through the right frontal lobe. A primary reason for that is:
A. It reduces the risk of bleeding complications
B. It reduces the risk of developing post-procedure epilepsy
C. It minimizes the risk of injuring eloquent brain tissue
D. It is the easiest site to access

Case 6 – Obstructive Hydrocephalus

The patient had developed obstructive hydrocephalus, as the presence of blood in the ventricles impaired the flow and drainage of spinal fluid. The pressure on the thalamus and brainstem result in obtundation. Untreated, progressive hydrocephalus is fatal. By placing an EVD and draining excess spinal fluid the hydrocephalus can be reduced and brain herniation prevented. Note that in some cases, patients do not recover the ability to drain spinal fluid and need a permanent, ventriculo-peritoneal shunt placed.

Teaching Point: The EVD is calibrated by aligning the drain with the ear (the external auditory meatus, or EAM), and CSF flow can be increased or decreased by changing this position.

1. What is the most likely reason why this patient developed hydrocephalus?
C. Obstruction of CSF drainage due to blood in the ventricles – the presence of blood in the ventricular system blocks the flow of CSF, resulting in its accumulation and development of hydrocephalus.

2. Which of the following symptoms are commonly caused by acute obstructive hydrocephalus?
A. Decreased level of consciousness – most commonly patients experience progressive lethargy and obtundation, although they may have focal signs as well.

3. EVD's are generally placed through the right frontal lobe. A primary reason is:
C. It minimizes the risk of injuring eloquent brain tissue – the non-dominant frontal lobe is chosen because injury in this area has the least risk of causing lasting deficits

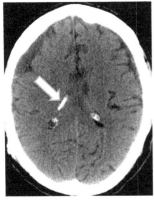

◀ *Non-enhanced head CT, showing significant reduction in the size of the lateral ventricles. The EVD catheter is visible (arrow).*

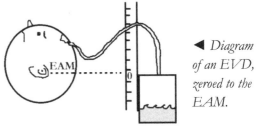

◀ *Diagram of an EVD, zeroed to the EAM.*

Case 7 – Neurosurgery

A 17 year old girl who has no past medical history was playing varsity basketball when she collided with another player and struck her head while attempting to rebound the ball. She fell to the ground with a brief loss of consciousness. After several seconds she awoke and was dazed but able to converse. She was evaluated on the sidelines and was found to have no retrograde or anterograde amnesia, no focal deficits, mildly impaired attention and headache. She was pulled from the game and evaluated in the ER where her exam was normal. A head CT was obtained which was negative.

One week later she presented to her primary care doctor with persistent headache, difficulty concentrating and sense of dizziness. She has no weakness, numbness or neck pain. Her physical exam is unremarkable.

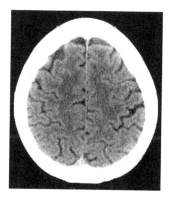

◀ *Unremarkable non-enhanced head CT, with no evidence of skull fracture or hemorrhage. Typical locations for traumatic hemorrhage include the anterior temporal lobes or orbitofrontal areas – places where the brain tissue could come into contact with bone during a high speed impact.*

1. When should this athlete return to play?
A. Once all symptoms have resolved
B. 72 hours after the loss of consciousness
C. Seven days after the loss of consciousness

2. Which of the following best describes this patient's lingering symptoms?
A. Typical post-concussive syndrome
B. Malingering
C. Moderate to severe traumatic brain injury

3. All of the following except which are common symptoms of concussion?
A. Disorientation (confused about date, time, location, etc.)
B. Vomiting
C. Personality or behavior changes
D. Brief seizure at the time of impact

Case 7 – Sports Related Concussion

This patient had a sports related concussion, a form of mild traumatic brain injury. Concussion is common, and nearly 1 in 10 high school athletes will have a concussion. Women have a higher risk of concussion, as do those playing soccer, football, rugby and hockey. A history of concussion or concussion with the past 10 days are also risk factors.

Symptoms of concussion can include brief loss of consciousness and temporary confusion, headache and dizziness, but these post-concussive symptoms typically resolve spontaneously within one to two weeks. Rarely, patients with concussion may have a more severe traumatic brain injury or spine injury. Head CT is not always needed unless a more severe TBI is suspected.

If you are a first responder to a sports related concussion you should start by addressing the ABC's – airway, breathing and circulation. If you suspect a neck injury or the patient remains unconscious, then do not move the head, neck or spine. A mental status exam including attention, memory and motor function should be performed, and the athlete should be observed for at least 3-4 hours. The patient should not return to play until fully evaluated and the symptoms have completely resolved.

Common Signs & Symptoms of Concussion

- Blank Stare
- Changes to balance, reaction time and coordination
- Disorientation
- Loss of consciousness – occurs in only about 10% of patients
- Memory loss of events before, during or after the injury
- Vomiting
- Emotional lability
- Slurred/unclear speech
- Headache and difficulty concentration
- Sensitivity to light/sound

1. When should this athlete return to play?
A. Once all symptoms have resolved – this is a clinical diagnosis

2. Which of the following best describes this patient's lingering symptoms?
A. Typical post-concussive syndrome, symptoms resolve spontaneously

3. All of the following except which are common symptoms of concussion?
D. Brief seizure at the time of impact – suggests a more severe TBI

Notes

Paul D. Johnson, MD

Image Atlas

"Me? I know who I am! I'm a dude playin' a dude disguised as another dude."
Tropic Thunder

Image Atlas: 1 – Perfusion Imaging

Perfusion imaging is a nuclear medicine technique that requires contrast, and provides information on cerebral blood flow. Perfusion imaging can be done with either CT or MRI. The most common use is to look for brain tissue that is experiencing decreased blood flow, but which hasn't yet died from severe ischemia – something known as 'penumbra' or 'tissue at risk.' Most recently, it has been shown that perfusion imaging can help select candidates for thrombectomy in the 6 – 24 hour time window.

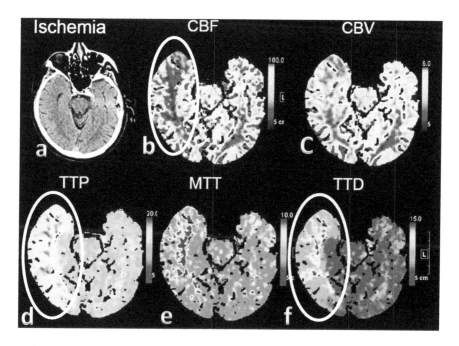

▲ *CT Perfusion study showing right temporal lobe ischemia. Although the non-enhanced CT (a) appears relatively normal, there are perfusion abnormalities (b-f). Although we won't go into detail here, each of the circled areas above demonstrates a different method of measuring total blood flow, blood volume, or delayed perfusion.*

Image Atlas: 2 – Fusiform Aneurysm

Fusiform aneurysms are elongated and dilated arteries, unlike the more common saccular (or "berry") aneurysm. They are more difficult to treat, as endovascular coils do not stay put.

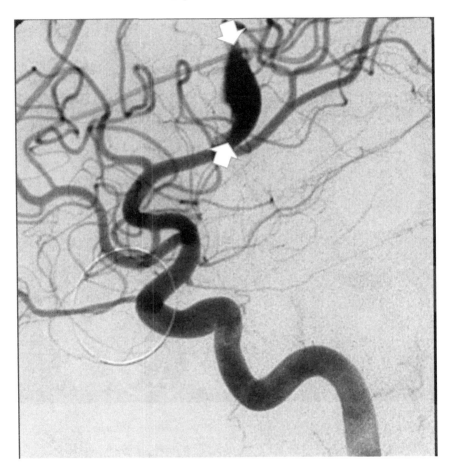

▲ *Conventional catheter angiogram of the internal carotid and its branches, demonstrating a typical, elongated fusiform aneurysm of a middle cerebral artery branch (between arrows).*

Image Atlas: 3 – Arachnoid Cyst

Arachnoid cysts are benign, space occupying lesions – that follow cerebrospinal fluid signal on all MRI sequences. They can become quite large, and although benign, can rarely cause symptoms due to mass effect. However, they are generally an incidental finding.

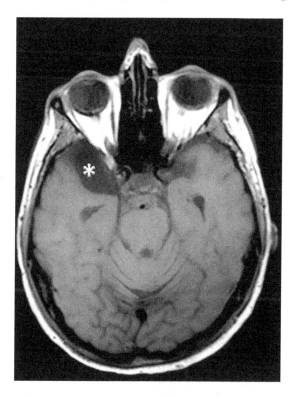

▲ *T1W brain MRI demonstrating a right temporal pole arachnoid cyst (*) – an incidental finding.*

Image Atlas: 4 – Dilated Perivascular Space

A perivascular space is the space between a penetrating blood vessel and the brain tissue – usually this is a microscopic area, but occasionally they can be quite large and can mimic a lacunar stroke. They are more common in patients with hypertension, and tend to occur in the basal ganglia and base of the brain most often. They are easier to identify on MRI than on CT.

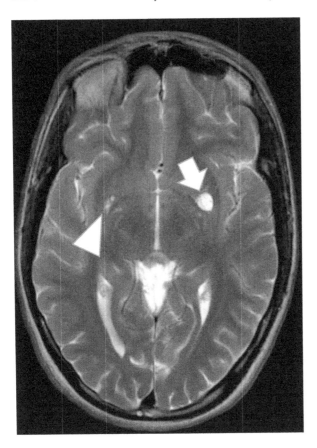

▲ *T2W brain MRI. There is a large dilated perivascular space in the left basal ganglia (arrow). Note the small blood vessel within that space. There are smaller dilated perivascular spaces on the opposite side (arrowhead).*

Image Atlas: 5 – Limbic Encephalitis

The limbic structures of the brain refers primarily to the hippocampus and other midline brain structures – these are some of the oldest parts of the brain, and are involved in memory, emotion and behavior. Some infections, such as Herpes Simplex Virus, as well as some auto-immune conditions, preferentially cause inflammation in the limbic system, which may lead to behavior changes and seizures.

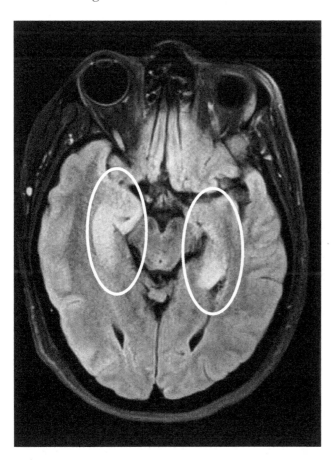

▲ *T2W FLAIR brain MRI, with inflammation in the bilateral hippocampus (encircled). Note that there may be some subtle difference in normal patients – often best to let the radiologist confirm the abnormality.*

Image Atlas: 6 – MR Spectroscopy

MR Spectroscopy attempts to identify the chemical structure of any tissue the radiologist selects during the scan. In theory it can help differentiate between tumor, infection and demyelinating lesions. In practice, it can be difficult to interpret and has several limitations – you're unlikely to see this obtained except in very rare circumstances.

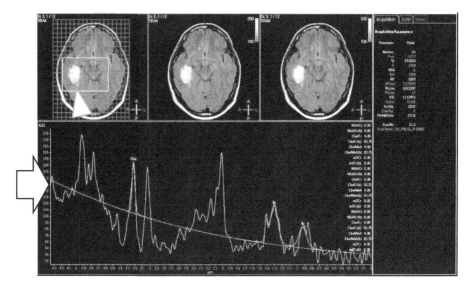

▲ *MR Spectroscopy. An area of interest is selected (arrowhead), and the chemical composition of the tissue is, in theory, displayed below (arrow).*

Image Atlas: 7 – Osmotic Demyelination Syndrome

Osmotic demyelination syndrome (ODS) occurs when there is rapid correction of long standing hyponatremia, and is the primary reason that hyponatremia is corrected slowly. The clinical and radiographic effects usually take several days to manifest. Clinically, the most common features are spastic paraplegia and dysarthria. The condition is also commonly seen following liver transplant. The condition can affect white matter anywhere, but is very common in the pons – it is sometimes referred to as Central Pontine Myelinolysis (CPM).

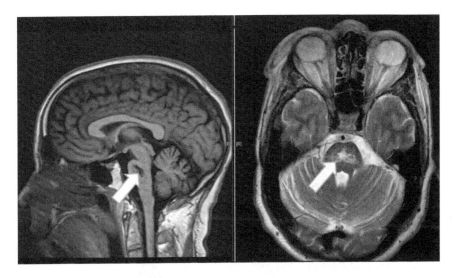

▲ *Sagittal T1W brain MRI (left) and axial T2W brain MRI (right) of the same patient.*

There is an area of encephalomalacia, or volume loss, in the pons which follows CSF signal on both sequences (arrows). This is a typical finding in central pontine myelinolysis, and represents loss of white matter tracts after correcting hyponatremia too quickly.

Image Atlas: 8 – Cavernous Malformation

A cavernous malformation, or cavernoma, is a vascular abnormality that is occasionally found in the brain or brainstem. These are dilated, sack like structures filled with slow flowing, stagnant blood. They can be small or large, stable or prone to bleeding. Usually they occur sporadically, but they frequently develop in patients who have had brain radiation. When there are multiple cavernomas, there may be an underlying genetic mutation, such as the CCM2 mutation. These are rarely treated, but may be resected if they are symptomatic.

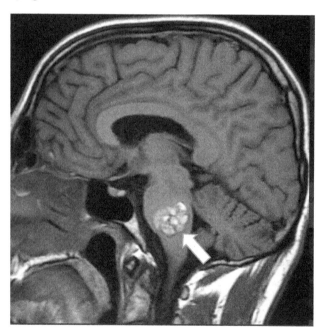

▲ *Sagittal T1W brain MRI with pontine cavernous malformation (arrow). The 'popcorn' like, lobulated appearance is typical. Because blood flow is so slow, these are not seen on angiography – MRI will typically show them best. When they hemorrhage, symptoms are usually milder than suggested by the size of the lesion.*

Image Atlas: 9 – Arteriovenous Malformation

An arteriovenous malformation, or AVM, is a vascular abnormality in which cerebral arteries have abnormal connections to cerebral veins, bypassing the capillaries and smaller arterioles. High pressure arterial blood flows directly in to thin walled veins, making these structures very prone to hemorrhage. In addition, due to the lack of capillaries brain tissue is not well perfused and is often dysfunctional. These are high risk for stroke, seizure and hemorrhage, and are often treated with either surgical resection or endovascular embolization.

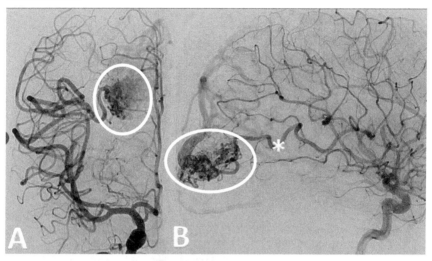

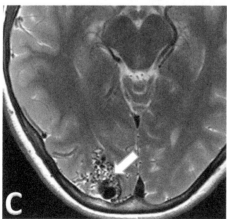

▲ *Coronal (a) and lateral (b) catheter angiogram showing an occipital AVM (circled). The feeding vessel is an MCA branch (*).*

◄ *T2W brain MRI (c) showing the flow voids for the same occipital pole AVM (arrow).*

Image Atlas: 10 – Developmental Venous Anomaly

A developmental venous anomaly (DVA) is a relatively common vascular abnormality, and benign. They are abnormally located veins, usually draining blood from a deep region of the brain and connecting directly to the brain surface. Surgical treatment of these is high risk, because they drain normal brain tissue, and removing one can lead to venous infarcts.

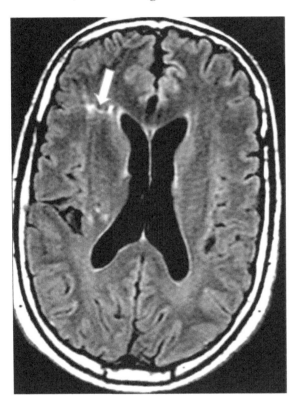

▲ *T2W FLAIR brain MRI showing a right frontal developmental venous anomaly (arrow), an incidental finding. These DVA's are simply large draining veins, but in abnormal locations. They may be associated with cavernous malformations.*

Image Atlas: 11 – Cavum Septum Pellucidum

An essentially benign finding – the thin 'septum pellucidum' separating the lateral ventricles may be duplicated.

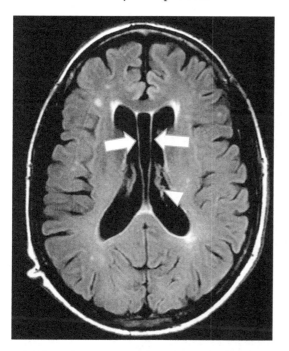

▲ *T2W FLAIR brain MRI showing a cavum (or duplicate) septum pellucidum, each of which is indicated by an arrow. This is benign, although there is some speculation that it is more common in people who have head traumatic brain injury. Note the prominent choroid plexus (arrowhead).*

Image Atlas: 12 – Mega Cisterna Magna

Another essentially benign finding – this represents an enlarged space within the posterior fossa, but is usually found incidentally. These patients generally have no symptoms, and no further evaluation or follow-up imaging is necessary.

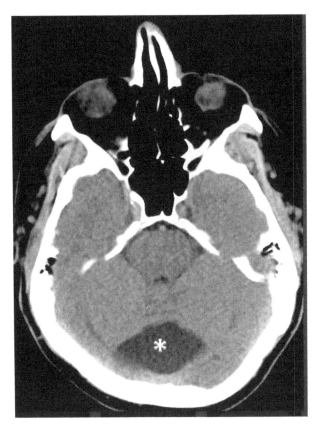

▲ *Non-enhanced head CT showing an incidental mega cisterna magna (*), a benign enlargement of the subarachnoid space.*

Image Atlas: 13 – Schizencephaly

Schizencephaly is a developmental anomaly of the brain, most commonly seen in pediatric neurology. There is failure of development of some part of the brain, leaving an open connection between the ventricles and subarachnoid space. There is no intervention.

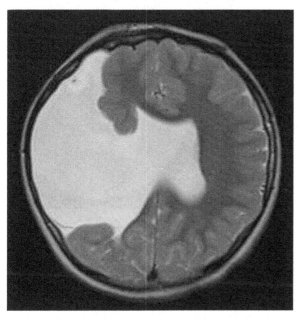

▲ *T2W brain MRI showing schizencephaly, with absent right middle cerebral artery territory brain tissue. This is a developmental anomaly.*

▼ *The difference between A) open lip and B) close lip schizencephaly. Note that the opening is lined with normal cortical tissue (i.e. grey matter).*

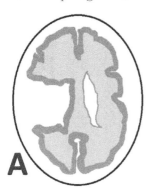
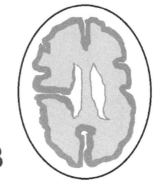

Image Atlas: 14 – Fahr Disease

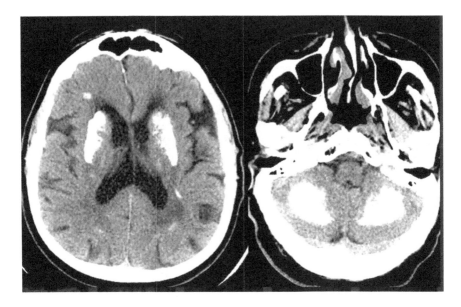

This patient has <u>extensive</u>, bilateral and symmetric deposition of calcium within the brain, consistent with Fahr syndrome. This can be due to a genetic condition – patients typically have depression and mild parkinsonism on exam.

However, it is also necessary to evaluate for disorders of calcium metabolism, including checking parathyroid levels.

Teaching Points: A small amount of symmetric, bilateral calcification in the basal ganglia is relatively common, and not necessarily a sign of disease – but this is abnormal!

The choroid plexus and pineal glands are common locations for 'normal' intracranial calcification.

Image Atlas: 15 – Perfusion Study for Brain Death

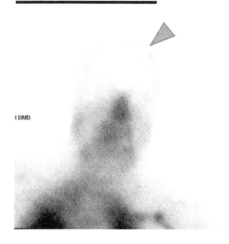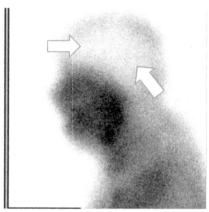

In a nuclear medicine perfusion study, areas with higher blood flow and metabolic activity are darker. Normally, the highly metabolic brain would be very dark – but in brain death, which is typically characterized by brain swelling, high intracranial pressure, and poor or absent cerebral blood flow, this signal is absent (arrows). Note the surrounding scalp (arrowhead) shows more blood perfusion than the brain.

Brain death evaluation follows strict legal criteria that vary state by state. However, nuclear medicine scans are used to confirm a clinical exam and history consistent with brain death in select cases. The brain death examination itself must only be done when there are no confounding factors, such as severe metabolic derangements or hypothermia, and in the setting of a brain injury that is consistent with the exam. In the US brain death requires absence of all cerebral and brain stem functions, including no spontaneous respirations and absent pupillary and gag responses.

Image Atlas: 16 – Syringomyelia

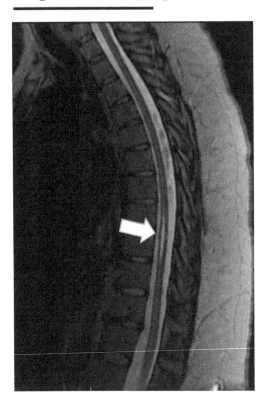

This sagittal T2W spine MRI shows syringomyelia (or syrinx), a fluid-filled dilatation in the center of the spinal cord. It often develops after a spinal cord injury, especially after trauma, inflammation such as infection or transverse myelitis, or due to altered CSF flow from spinal tumors. They may be symptomatic or asymptomatic. Also note the association with a Chiari malformation, as seen on page 226.

Image Atlas: 17 – Brain herniation

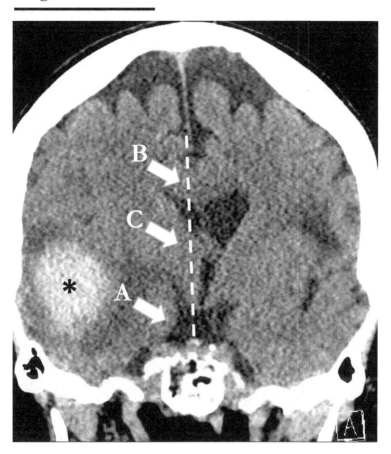

This coronal non-enhanced head CT demonstrates a large right temporal lobe intraparenchymal hemorrhage (*) resulting in significant mass effect and brain herniation (arrows). A) Transtentorial herniation of the uncas of the temporal lobe may compress CN III, resulting in an ipsilateral blown pupil. B) Subfalcine herniation can result in anterior cerebral artery compression. C) Pressure on the thalamus and brainstem can result in coma and death. Osmotic therapies, such as hypertonic saline or mannitol, can be used to reduce brain edema and prevent herniation. Usually these measures are temporary and serve as a bridge to surgical intervention.

Image Atlas: 18 – Frontotemporal Dementia (FTD)

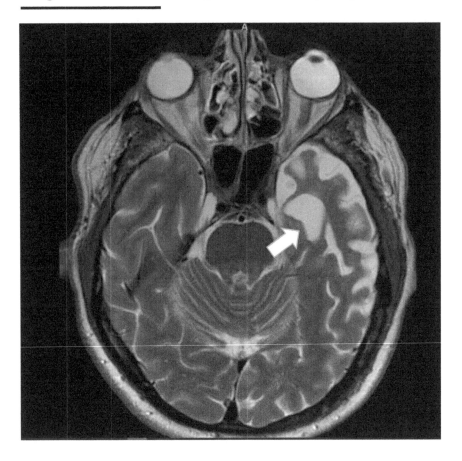

The frontotemporal dementias (FTD) represent a diverse group of neurodegenerative diseases, with different underlying pathologic causes and different clinical presentations. They are characterized by relatively rapid progression (months to years), and lobar degeneration on brain imaging. FTD is a common cause of dementia in the young, typically presenting before age 60.

Common presentations include the **behavioral variant**, marked by focal frontal and temporal lobe atrophy and a marked inhibition of behavior, including gloss of sympathy and apathy. Another common form is **primary progressive aphasia**, characterized by a marked alteration in language. The case shown here, with left temporal lobe predominant atrophy (arrow), presented with a dramatic paucity of vocabulary.

Image Atlas: 19 – Subclavian Steal

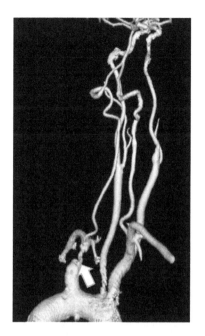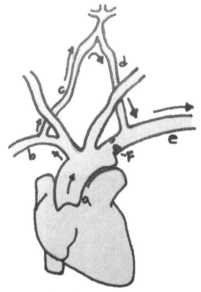

Subclavian steal is a rare phenomenon in which blood flow to the posterior circulation is diverted to supply one of the limbs, resulting in presyncopal episodes, usually triggered by prolonged use of the affected limb and resolving with 10-15 minutes of rest. In order for this to occur there must be a proximal, tight stenosis in one of the subclavian arteries (arrow in the computer reformatted CT angiogram, letter "F" in the diagram). Blood flows normally through the aorta (a) through the normal subclavian artery (b), up the normal vertebral artery (c), then retrograde down the vertebral artery on the affected side (d) to supply the diseased subclavian artery (e).

Image Atlas: 20 – Tuberous sclerosis complex

Tuberous sclerosis is a genetic condition leading to widespread hamartomas throughout the body – mainly in the brain, heart, kidney, liver, lungs, skin and eyes. Within the brain, subependymal giant-cell astrocytomas (SEGA, arrow on image A) are common, as are cortical tubers (arrows, image B). Although cardiac rhabdomyoma is the primary cause of mortality and leading reason why these children come to medical attention, central nervous system complications are also very common. Many children develop epilepsy, cognitive impairment or other behavioral problems.

The primary genes are TSC1 and TSC2, which encode the proteins hamartin and tuberin, respectively.

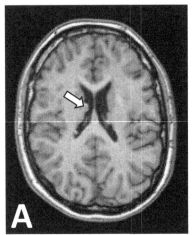
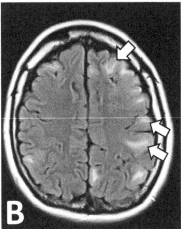

Image Atlas: 21 – CADASIL

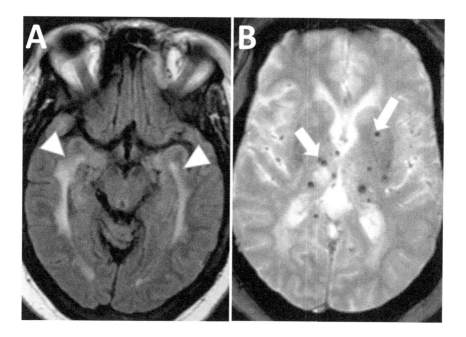

This MRI is from a patient with CADASIL (Cerebral Autosomal Dominant Arteriopathy with Subcortical Infarcts and Leukoencephalopathy). This is hereditary stroke syndrome is due to a mutation of the Notch 3 gene on chromosome 19. Patients frequently develop migraine with aura and behavioral issues as young adults, followed by lacunar strokes and extensive white matter disease. Imaging findings suggestive of CADASIL include white matter disease extending into the temporal poles (image A, arrowheads) as well as white matter disease of the extreme capsule (not pictured). These patients also often have extensive micro-hemorrhages on GRE (image B, arrows), which puts them at risk for bleeding. Typical management includes treating all modifiable stroke risk factors, such as blood pressure, cholesterol and tobacco use. The clinical impact of this disease is variable, but can lead to multi-infarct dementia.

Notes

CASE INDEX

Neuroimaging Cases
Epidural hematoma…..59
Infarct on MRI…..61
Subarachnoid hemorrhage…..63
Neuroanatomy on MRI…..65
Infarct on CT…..67
Infarct on MRI…..69
Benign intracranial calcification…..71

Stroke
Transient Ischemic Attack…..81
Carotid artery stenosis…..83
Intraparenchymal hemorrhage, hypertensive…..87
Large vessel occlusion…..91
Giant cell arteritis……95
Lateral Medullary Stroke…..97

Epilepsy
First seizure…..105
Epilepsy in pregnancy…..107
Periodic epileptiform discharges (PLEDS) …..109
Antiepileptic Drug Quiz…..111
Non-epileptic spells…..113
Acute symptomatic seizures…..115
Status epilepticus…..117
Temporal lobe epilepsy…..123

Headache
Migraine…..129
Idiopathic intracranial hypertension…..135
Cluster headache, other trigeminal autonomic neuralgias…..137
Headache in pregnancy…..139
Dural venous sinus thrombosis…..141
Posterior reversible encephalopathy syndrome…..143
Meningitis…..145

Sleep
Obstructive sleep apnea…..151
Circadian rhythm disorders…..153
Narcolepsy with cataplexy…..155
Restless leg syndrome…..157

Movement
Parkinson disease…..163, 169
Essential tremor…..165
Parkinson disease medications…..167
Hepatocerebral degeneration…..171
Progressive Supranuclear palsy…..173
Huntington disease…..177
Acute dystonic crisis…..179
Lewy Body dementia…..181

Neuromuscular
Myasthenia gravis…..187
Charcot-Marie-Tooth disease (HSMN) …..189
Inclusion body myositis…..191
Guillan-Barré Syndrome…..193
Amyotrophic lateral sclerosis…..195
Duchenne muscular dystrophy…..197
Peripheral neuropathy…..199
Subacute combined degeneration…..203
Medication induced myopathy…..205
Lambert-Eaton myasthenic syndrome…..207

Dementia
Creutzfeldt-Jakob disease…..217
Mild cognitive impairment…..219
Alzheimer dementia…..221
Delirium…..223
Transient global amnesia…..225
Normal-Pressure Hydrocephalus…..227

Immunology
Optic neuritis…..233

Multiple sclerosis.....235

Behavioral Health
Borderline personality disorder.....243
Psychiatric medication side effects.....245
Depression.....247
Defense mechanisms.....249
Schizophrenia.....251
Withdrawal syndromes quiz.....253

Neurosurgery
Subarachnoid hemorrhage, aneurysmal.....257
Chiari 1 malformation.....263
Diffuse axonal injury.....265
Glioblastoma Multiforme.....267
Cervical spondylotic myelopathy.....273
Acute obstructive hydrocephalus.....275
Sports related concussion......277

Imaging Atlas
Perfusion imaging.....281
Fusiform aneurysm.....282
Arachnoid cyst.....283
Dilated perivascular space.....284
Limbic encephalitis.....285
MR spectroscopy.....286
Osmotic demyelination syndrome.....287
Cavernous malformation.....288
Arteriovenous malformation.....289
Developmental venous anomaly.....290
Cavum septum pellucidum.....291
Mega cisterna magna.....292
Schizencephaly.....293
Fahr disease.....294
Perfusion study for brain death.....295
Syringomyelia.....296
Brain herniation.....297
Frontotemporal dementia.....298
Subclavian steal syndrome.....299

Tuberous sclerosis complex…..300
CADASIL…..301

INDEX

ABCD2 score.....78
Absence seizures.....121
Afferent pupillary defect.....32, 234
Apnea-Hypopnea Index.....152
AIDP.....193
Akathisia.....176
Altitudinal visual loss.....96
Alzheimer disease.....221
Amantadine.....167
Amyotrophic lateral sclerosis.....195
Aneurysm treatment.....259
Anisocoria.....33
Antiepileptic drug levels.....111
Antiepileptic medication quiz.....111
Antiepileptic medications.....104
Aphasia flow chart.....31
Arachnoid cysts.....283
Arteriovenous malformation.....289
Autonomic neuropathy.....200
Axon.....186

B-12 deficiency.....204
Babinski sign.....42
Basal ganglia.....161
Becker muscular dystrophy.....198
Bell's palsy.....36
Benign neonatal myoclonus.....150
Benign paroxysmal positional vertigo (BPPV).....34
Benign rolandic epilepsy.....121
Birth control in epilepsy.....108
Blepharospasm.....180
Botox, for migraine.....132
Brachial plexus.....210
Brain death.....295
Brain tumors.....270
Brainstem.....7
Breastfeeding in epilepsy.....107
Broca's area.....6
Bruxism.....150

CADASIL.....301
Carbamazepine.....104
Carbidopa/levodopa.....167
Carotid endarterectomy.....85
Carotid stenosis.....85
Carpal tunnel syndrome.....39
Catheter angiography.....53

Cavernous malformation.....288
Cavum septum pellucidum.....291
Central nervous system.....2
Central pontine myelinolysis.....287
Central sleep apnea.....152
Cervical myelopathy.....273
Charcot-Marie Tooth.....189
Chiari malformation.....263
Cholinesterase inhibitors.....220
Chorea.....175
CIDP.....194
Circadian rhythm disorders.....153
Circle of Willis.....21
Cluster headache.....137
COMT.....167
Concussion.....277
Coordination exam.....41
Copaxone.....238
Corticobasal degeneration.....174
Cranial nerves.....8
Craniopharyngioma.....271
Creutzfeldt-Jacob disease.....217
CSF dynamics.....228
CSF normal labs.....25
CT angiography.....49
CT scans.....45
Cytoalbuminologic dissociation.....194

DaT scan.....165
DAWN.....94
Deep brain stimulator.....169
Defense mechanisms.....249
DEFUSE 3.....94
Delayed sleep phase syndrome.....154
Delirium.....223
Dementia.....215, 220
Depression.....247
Dermatomes.....211
Dermatomyositis.....192
Developmental venous anomaly.....290
Diffuse axonal injury.....265
Dilated perivascular space.....284
Dimethyl fumarate.....238
Dopamine agonists.....167
Dopamine.....162
Dravet syndrome.....121
Duchenne muscular dystrophy.....197
Dural venous sinus thrombosis.....141

DWI.....52
Dysarthria.....36
Dystonia.....176
Dystonic crises.....179

EEG.....101, 122
EMG/NCS.....185
Empty delta sign.....142
Ependymoma.....271
Epidural hematoma.....59
Epilepsy, causes of.....116
Epilepsy, medically refractory.....120
Epilepsy, pediatric syndromes.....121
Epilepsy.....101
Epworth sleepiness scale.....149
Essential tremor.....165
EVD.....275
Extracranial circulation.....19
Extra-ocular movements.....10, 33

Facial nerve.....35
Fahr disease.....295
FDG-PET.....222
Febrile seizures.....121
Fingolimod.....238
First seizure.....105
FLAIR.....51
Fosphenytoin.....104
Frontotemporal dementia.....298
Fusiform aneurysm.....282

GABA.....162
Gait exam.....42
Gelastic seizures.....123
Giant cell arteritis.....95
Glatiramer acetate.....238
Glioblastoma.....267
Glutamate.....162
Guillan-Barre syndrome.....193

Headache during pregnancy.....139
Headache red flags.....130
Headache.....127
Hemiballismus.....175
Hemicrania continua.....138
Hemiparesis.....82
Hemiplegia.....82
Hemorrhagic stroke.....87, 90
Hepatocerebral degeneration.....171
Herniation.....297
Hippocampal ablation.....120
Hippocampus.....241
HMSN.....189

Homunculus.....6
HSV meningitis.....145, 285
Hunt & Hess score.....258
Huntington disease.....177
Hydrocephalus.....275
Hyper-ammonemia.....171
Hypercoagulable labs.....80
Hypertonic saline.....297

Idiopathic intracranial
 hypertension.....135
Inclusion body myositis.....191
Insomnia, treatment.....154
Interferon.....238
Internuclear ophthalmoplegia.....236
Intracranial circulation.....20
It be iddy biddy baby doo doo.....56
IVIG.....194

Juvenile myoclonic epilepsy.....121

Lacosamide.....104
Lambert-Eaton myasthenic
 syndrome.....207
Lamotrigine.....104
Language exam.....30
Large vessel occlusion.....91
Lateral medullary syndrome.....97
Lennox-Gastaut syndrome.....121
Levetiracetam.....104
Lewy-Body dementia.....181
Limbic encephalitis.....285
Limbic system.....241
Lithium.....245
Lobes of the brain.....4
Lower motor neuron.....3
Lumbar plexus.....211
Lumbar puncture.....24

Mannitol.....297
MAO-B Inhibitor.....167
McDonald criteria.....232
Median nerve.....39
Medication induced myopathy.....206
Medication overuse headache.....134
Medulloblastoma.....272
Mega cisterna magnum.....292
Meningioma.....270
Meningitis.....145
Mental status exam.....28
Meralgia paresthetica.....209
Migraine and stroke.....127
Migraine aura.....133

Migraine, acute treatment…..131
Migraine, chronic treatment…..132
Migraine…..129
Mild cognitive impairment…..219
Mini-mental exam…..29
Montreal cognitive exam…..216
Motor cortex…..5
Motor exam…..37
Motor neurons…..2
Motor unit…..185
MPTP…..162
MR spectroscopy…..286
MR angiogrpahy…..53
MRC muscle scale…..37
MRI…..49
Multiple sclerosis, CSF…..238
Multiple sclerosis…..231, 235
Multiple sleep latency test…..149
Multi-system atrophy…..174
Myasthenia gravis…..187
Myelin…..3
Myelopathy…..204, 273
Myoclonus…..175

Narcolepsy with cataplexy…..155
Natalizumab…..238
NBME…..ii
Neglect…..38
Neuroleptic malignant syndrome…..251
Neuromuscular quick reference…..209, 212
Neuromyelitis optica…..2237
Neurotransmitters…..162
Night terrors…..150
Nightmares…..150
NIHSS…..76
Non-arteritic ischemic optic neuropathy…..96
Non-epileptic spells…..114
Normal-pressure hydrocephalus…..227
NOTCH 3…..301
Nuclear medicine scan…..295

Obstructive sleep apnea…..151
Ocrelizumab…..138
Oculogyric crises…..180
Oligodendroglioma…..272
Opisthotonos…..180
Optic nerve exam…..32
Optic neuritis…..233
Osmotic demyelination…..287

Painful peripheral neuropathy…..202

Papez-circuit…..241
Parasomnias…..150
Parkinson disease…..163, 169
Parkinsons plus syndromes…..174
Paroxysmal hemicrania…..138
Perfusion imaging…..281
Peripheral nervous system…..2
Peripheral neuropathy…..199
Peroneal nerve…..41
Personality disorders…..243
Phenobarbital…..104
Phenytoin…..104
PHQ-2…..247
PLEDs…..109
PLMD…..150
Polymyositis…..192
Polysomnogram…..149
Porencephaly…..294
Pregabalin…..104
Pregnancy in epilepsy…..108
PRES…..143
Primary progressive aphasia…..298
Prion disease…..217
Progressive supranuclear palsy…..173
Pronator drift…..38
Pseudo-flair, MS…..236
Pseudotumor cerebri…..see Idiopathic intracranial hypertension

Radial nerve…..40
Rasmussen encephalitis…..121
Rathke cleft cyst…..271
Reflexes…..41
Restless leg …..157
Ring enhancing lesions…..269
Rituximab…..238
RT-Quic…..218

SAH, complications…..261
SAH, Ottawa rule…..258
Schizencephaly…..293
Schizophrenia…..251
Schwannoma…..270
SEGA…..300
Seizure classification…..103
Seizure precautions…..106
Sensory exam…..38
SIG E CAPS…..248
Sleep hygiene…..158
Sleep walking…..150
Sleep…..149
Small fiber neuropathy…..202
Somatosensory cortex…..5

Spinal cord anatomy.....22
Spondylotic myelopathy.....273
Statin myopathy.....205
Status epilepticus.....118
Stereotypy.....176
Streak artifact.....46
Stroke.....75
Subacute combined degeneration.....203
Subarachnoid hemorrhage.....257
Subarachnoid hemorrhage.....63
Subclavian steal.....299
SUNCT.....138
Syrinx.....264, 296

T1W.....50
T2W*.....52
T2W.....50
Tardive dyskinesia.....176
TCD (Transcranial Doppler).....54
Tecfidera.....238
Temporal lobe epilepsy.....123
Tension type headache.....128
Thymoma, in myasthenia.....188
TIA.....76, 81
Tibial nerve.....41
TIPS.....171
Topiramate.....104
Torticollis.....180
Toxoplasmosis.....175
tPA.....83, 86
Transient global amnesia.....225
Transverse myelitis.....237
Trigeminal autonomic cephalgia.....138
Tuberous sclerosis complex.....300

Ulnar nerve.....40
Ultrasound.....54
Upper motor neuron.....3

Vagal nerve stimulator.....120
Valproate.....104
Venography.....56
Ventriculoperitoneal shunt.....275
Ventriculostomy.....275
Visual fields.....11, 34

Wallenberg syndrome.....97
Wernicke's area.....6
WFNS.....258
Withdrawal syndromes.....253

Zonisamide.....104

Notes

ABOUT THE AUTHOR

Paul D. Johnson, MD is a fellowship trained vascular neurologist and medical director of a comprehensive stroke center in the mountain west.

For questions or errata, send email to Crow.Neurology@gmail.com

Some images were purchased from ShutterStock.com

Made in the USA
Middletown, DE
25 September 2023

39369556R00179